Small Dams

Small Dams

Planning, Construction and Maintenance

Barry Lewis
Dam Consultant, Melbourne, Australia

CRC Press
Taylor & Francis Group
Boca Raton London New York Leiden

CRC Press is an imprint of the
Taylor & Francis Group, an **informa** business

A BALKEMA BOOK

CRC Press/Balkema is an imprint of the Taylor & Francis Group, an informa business

© 2014 Taylor & Francis Group, London, UK

Typeset by MPS Limited, Chennai, India

Library of Congress Cataloging-in-Publication Data

Lewis, Barry, 1942–
    Small Dams : Planning, Construction and Maintenance / Barry Lewis, Dam Consultant, Melbourne, Australia.
        pages cm
    Includes bibliographical references and index.
    ISBN 978-0-415-62111-3 (hardback)
    1. Reservoirs—Design and construction.   2. Water-supply, Agricultural.   I. Title.
    TC540.L55   2013
    627'.8—dc23
                                                                            2013022561

Published by: CRC Press/Balkema
                P.O. Box 11320, 2301 EH, Leiden, The Netherlands
                e-mail: Pub.NL@taylorandfrancis.com
                www.crcpress.com – www.taylorandfrancis.com

ISBN: 978-0-415-62111-3 (Hbk)
ISBN: 978-1-315-85774-9 (eBook PDF)

Disclaimer

This book is intended for use as a guide to owners and operators of small dams. It suggests prudent approaches to normal surveillance and maintenance practice with a view to enhancing the long-term safety and survival of small dams. It is not intended as a source of detailed information to cover all possible eventualities. In the event of any suspected imminent or potential failure condition, expert advice should be sought immediately.

Table of contents

Preface

Farmers are well aware of the need to boost productivity. In the face of greater competition for domestic and overseas markets, the farmer who wants to succeed has to take a business person's approach to increasing efficiency, reducing costs and improving output. In this environment, water becomes an economic factor and its provision a matter for careful deliberation.

This book is designed as a guide for small dam owners, engineering students, Government agencies, developers, and earthmoving contractors who are responsible for designing, building and using the majority of water storages constructed. It is also designed for engineers who have not specialised in small earth dam design for agricultural hydrology, but who may be called upon, from time to time, to design small water storage schemes. It is not intended to replace standard procedures currently used by those specialised engineers who are engaged in farm water design, although many of the design methods described herein are based on their procedures. It does, however, attempt to provide such engineers with a comprehensive array of design data and a concise reference to basic design techniques that are not otherwise readily available.

To cover all aspects of water conservation and use in detail is not possible in a book of this size. However, it will give the landowner an insight to those aspects of planning which must precede the establishment of a feasible and economic water supply project.

The information for this book has come from a number of sources. One is a series of small dam pamphlets (which I wrote for the Rural Water Authorities in Australia over a number of years), another is data from papers that I have read and presented at conferences around the world.

Barry Lewis
May, 2013

Acknowledgments

There are many people who have supported and provided information, ideas and procedures in the collation of this book on small dams. Perhaps a major dilemma for me is the incorporation of sources of information, which are better explained in the anonymous quote:

> *How about the many ideas and procedures that one picks up from discussion with colleagues? After the passage of time, one can no longer remember who originated what idea. After the passage of even more time, it seems to me that all of the really good ideas originated with me, a proposition that I know is indefensible.*

It becomes clearer with time that nothing is new, but people forget the original sources and whether those sources were really original or were the result of slight modifications of other people's ideas. It is like the wheel. Nobody knows who conceived the idea, yet it is used universally. So, credit must be given to those who have gone before me in this area.

My family has been supportive in assisting in many ways. My wife, Marion, has typed and corrected a number of drafts. Richard, my son, has shown me some of the intricacies of the computer, whilst Janette, Helen and Kenneth have done their bit in other ways.

In the practical areas of planning, design, construction, maintenance and legal issues of small dams and other related areas I would like to thank all of those who have been instrumental in the completion of this book.

Introduction

People have always gathered water during wet seasons so as to have enough for themselves, their animals and their crops in dry spells. The earliest known dams were in China in the sixth century BC. The ruins of ancient dams also exist in the Tigris and the Nile River Valleys. Some Roman dams built in Italy, Spain and North Africa are still being used today.

Today, dams are built to allow storage of water to give a controlled supply for domestic or industrial consumption, for irrigation, to generate hydro-electric power, or to prevent flooding. Large dams are built of earth, rock, concrete or a combination of these materials (for example, earth and rock fill). They are built as: gravity dams, where the stability is due entirely to the great weight of material; arch dams, where abutments at either side support the structure; or, arch gravity dams, which are a combination of the two.

Those who plan, design, construct, maintain, use and administer significant infrastructure developments which have a real potential to harm people and property if something goes wrong, are potentially subject to significant legal liabilities. These liabilities need to be taken into account when decisions are made in relation to such developments.

It is well known that Australia is a dry continent characterised by variable rainfall. It is less well known that, in response to widespread harvesting of water on the small and large scale, Australia has the highest water storage per capita in the world (Lewis and Perera, 1997; Lewis, 2001b). Small dam development has occurred in response to agricultural expansion, and to the need for a reliable source of water for stock, domestic and irrigation use, particularly during periods of drought. However, there is a growing belief in the community that small dams are impacting on water resources in many major catchments by reducing stream flow and flow duration. Potential and actual impacts of such reductions on the conflicting needs of the environment, agriculture and industry are cause for concern. These concerns are set against a background of changes including increased areas of intensive land uses such as viticulture and horticulture, and the development of farmland into rural residential subdivisions in commuter belts surrounding major cities and rural centres. These changes are associated with increased small dam development. In a significant proportion of cases, particularly where intensive land use changes have occurred, small dams are constructed that exceed the available water resource. Where this occurs, downstream impacts on stream flow will be the cause of conflict between users (including the environment). In addition, landowners may experience financial losses by constructing dams of a size inappropriate for the catchment.

In most countries around the world and in Australian States, provisions of the relevant *Water Act*(s) require that environmental consideration be taken into account in relation to new and renewed licences. These requirements apply to both regulated and unregulated waterway systems. Since unregulated waterways may not have large storage capacities from which environmental releases can be made, flow needs to be specified as a proportion of daily stream flow, or as a minimum daily flow. Many factors such as in-stream habitat requirements need to be considered when allocating an entitlement to environmental flows. In the case of regulated waterways, environmental flows can be released as part of the normal operating procedure.

Planning

1.1 ASSESSING WATER NEEDS

Water is the most plentiful and vital liquid on earth. No life, as we know it, could exist without water as it provides the essential medium in which all things grow and multiply.

In recent years the steady growth in the numbers of agricultural businesses, including viticulture, aquaculture, irrigating crops and grazing animals, has increased the demand for water supplies. No longer can the small, muddy small dam or the small bore with windmill provide adequate supplies of clean water for these uses.

Landholders are lucky if their property can be serviced from a local water trust or authority. In most cases, the only option is to utilise on-farm water resources and this requires careful planning prior to costly construction. In a majority of cases farmers use roofs of buildings to collect rainwater for domestic and farm purposes. Rainwater is the main source of high quality water for human consumption, providing that guttering and tanks are kept in good order. In some cases underground water is used, but it may require treatment to remove or neutralise the minerals it contains. Where the terrain permits, rainfall and run-off gravitating to various categories of dams (for example, off-stream and on-stream catchment dams), provide the farm water supplies.

1.1.1 Planning water supplies

When deciding the most appropriate source of water for a farm, the following steps should be taken:

- determine the purpose for which the water is to be used (for example, household, garden, irrigation, stock, aquaculture, viticulture or a combination of all of the above);
- determine the quantity of water required for the different purposes, time of year and seasonal use and conditions (for example, summer, winter, flood or drought);
- ascertain whether the source of water supply is regulated (controlled from a reservoir that releases flows subject to availability) or unregulated (for example, bore water, rainfall or combination of both) and can be maintained when supplies are needed.

These factors are a guide to planning and they differ with almost every property. The most likely base can be established by calculating the average daily or average annual requirements for each proposed use of the dam. It is only after these facts are established that harvesting, storing and distributing the water resource can be planned (Lewis, 1993).

1.1.2 Water quality

In project planning the quality of water can determine success or failure of a small dam for several reasons. First, the quality of water available from a farm water system will determine the use to which that water can be put, and hence govern the overall feasibility of the system. Many natural waters have impurities that can make them directly harmful to crops. Therefore, knowledge of the quality of the water supply is essential.

Second, water quality is a factor in determining storage capacity. For example, if saline water is intended for irrigation use, additional quantities of non-saline water must be applied from time to time to avoid damage to the irrigated crops or soils.

Third, the presence of certain mineral constituents (such as calcium, potassium, magnesium and/or sodium cations) in stored water may cause tunnel erosion in some soil types. Therefore, the quality of water to be stored in a small dam must be considered in the design and construction of the impounding embankment.

Finally, overland flows mobilise and transport nutrients, fertilisers and silt which can contaminate the dam water supply. Precautions need to be taken to minimise the risk of this occurring, as algal blooms and weed infestation could result (see Section 7.4.1).

When planning for farm water supply, the presence of minerals, sediments, nutrients, agricultural chemicals, biota or bacteriological contamination needs to be considered, according to whether the water is to be used for:

i Water for consumption for humans and animals

- Micro-organisms and other harmful organisms which may produce illness or disease.
- Soluble minerals and other substances which give the water an unpleasant taste or make a person unwell.
- Silts, suspended impurities or algae which may cause discolouration, odour or taste problems and in some cases are toxic to both humans and animals.

ii Water for household, dairy or general farm use

- Suspended minerals and compounds which cause problems with clothes washing and the build-up of scale in equipment such as pipes, hot water systems and cooking utensils.
- Impurities which can cause water discolouration and associated staining and odour problems.
- Acids, dissolved minerals and gases which corrode pipes and equipment.

iii Water for irrigation

- Dissolved minerals which can alter water and soil chemical and physical properties. For example, an altered pH may inhibit plant growth.

iv Water to be stored behind small dams

- Dissolved and suspended particles can affect the soil structure of a dam, which can lead to tunnel erosion.

The testing of water samples to determine the presence and likely effect of these impurities requires laboratory facilities (see Section 7.2).

1.2 ASSESSMENT OF CATCHMENT YIELD

The initial question to be asked when considering the development of any storage dam on a property is how much water can be harvested in a catchment from overland run-off.

1.2.1 Factors controlling catchment yield

Estimating catchment yield by the methods described needs to be done within the context of a full understanding of the run-off process, so that the limitations of the method can be appreciated. Catchment yield is the volume of water that flows from a catchment past a given point (such as a stream gauging station) and is generally calculated on an annual basis (Beavis and Lewis, 1999). It comprises surface run-off and base flow (discharge from shallow and deep groundwater). Catchment yield will vary according to a number of hydrological and physical factors, which control how much water is delivered to, retained by and transported from a catchment.

i Hydrological factors

Run-off may be regarded as the residue of rainfall after losses due to interception by vegetation, surface storage, infiltration, surface detention and waterway detention. There is always a time lag between the beginning of rainfall and the generation of run-off. During this time lag, rainfall is intercepted by vegetation, infiltrates the soil, and surface depressions start to be filled. Run-off occurs when the infiltration capacity is exceeded, or the precipitation rate exceeds the rate of infiltration. Thus the depth of water builds up on the surface until the head is sufficient to result in run-off. As the flow moves into defined waterways, there is a similar build-up in the head with volume of water, and this is termed waterway detention. The water in surface storage is eventually diverted into infiltration or evaporation pathways (Beavis and Howden, 1996; Beavis and Lewis, 1999).

Rainfall duration, intensity, and distribution influence both the rate and volume of run-off. Infiltration capacity normally decreases with time, so that a short storm may not produce run-off, in contrast to a storm of the same intensity but of long duration. However, an intense storm exceeds the infiltration capacity by a greater margin than a gentle rain. Therefore, the total volume of run-off is greater for an intense storm even though total precipitation is the same for the two events. In addition, an intense storm may actually decrease the infiltration rate by its destructive action on the structure of the soil surface. Generally, the maximum rate and volume of run-off occur when the entire catchment contributes. However, an intense storm on one portion of the catchment may result in a greater run-off than a moderate storm over the entire catchment.

ii Physical factors

Physical factors affecting run-off include topography, soil type and antecedent soil moisture conditions, catchment size, shape and orientation, and management practice. Topographic features such as slope and the extent and number of depressed areas influence the volume and rate of run-off. Catchments having extensive flat or depressed areas without surface outlets have lower run-off than areas with steep, well-defined drainage patterns. A catchment with a northerly aspect will dry out faster than one with a southerly aspect. Soil texture, fabric, clay mineralogy and antecedent moisture conditions have major impacts on the infiltration rate and capacity, and thus influence run-off. Sandy soils with an open fabric have high infiltration rates and generate less run-off than clayey soils with a closed fabric (where intergranular spaces are smaller and there is less connectivity between pores). Both run-off volumes and rates increase with catchment area. However, run-off rate and volume per unit area decrease as the area increases. Vegetation cover and management practices influence infiltration rate by intercepting precipitation and modifying soil structure respectively. Vegetation also retards overland flow and increases surface detention to reduce peak run-off rates. Structural works such as dams, weirs, pipe culverts and levees all influence run-off rates by either directing surface water into preferential flow lines or storing water.

1.2.2 Methods of estimating catchment yield

The most readily available source of water is the surface water in rivers and lakes. This water is usually stored in dams. In certain parts of Australia, fortunate farmers have 'run of the river' schemes, that is, they do not need storages because the flows in the rivers are so reliable that they can meet all requirements. This is the situation in areas of consistently high rainfall, an uncommon circumstance in Australia. It also applies to farms that draw their water from a river downstream from a large public storage.

In the absence of a reliable flow in the river, a storage must be constructed to meet all the water requirements for the storage period as well as evaporation and seepage losses. The storage period is that interval during which there is no run-off into the dam. Table 1.1 gives a 'rule-of-thumb' estimate of the storage period required in areas with different average rainfall.

Run-off from a catchment must be adequate to meet requirements. In practice, the catchment should be neither too small nor too large. If it is too small, there are two dangers. First, it will not meet the requirements, and second, the lack of silt, which all streams carry, will result in excessive seepage from the dam. All earth dams seep to a greater or lesser degree. However, after a couple of years, the silt from the stream impregnates the floor and this forms a relatively watertight seal.

Table 1.1 Storage period.

Average annual rainfall (mm)	Storage period (months)
>650	12
450 to 650	18
250 to 449	24
<250	30–36

Source: Modified from Soil Conservation Authority, 1983.

If the catchment is too large, the dam will be subject to very heavy floods that, in turn, necessitate a concrete spillway. This is an expensive item and frequently makes a project uneconomical.

Two methods are currently used in Australia for estimating the yield of a catchment. One is the United States Department of Agriculture (USDA, 1969) method of estimating daily run-off. This is based on daily rainfall records for the district. The longer the period of record, the better the results. The other method is based on the assumption that catchment yield is a percentage of the average annual rainfall. Variability of rainfall in Australia limits the accuracy of forecasting, and hence, the reliability of this method. It therefore follows that, despite the most careful calculations, it is difficult to guarantee that a dam will always meet requirements. However, a method of estimating the potential catchment yield must be adopted so that a farm water supply scheme can be planned on a reasonably sound basis. The run-off yield method was first published by the Water Research Foundation of Australia (Burton, 1965) and has proved to be of continued practical value in water resource management in most States. Table 1.2 provides an estimate of yields from small natural catchments. The reliability column relates to the number of years in a ten-year period in which the given percentage yield will be equalled or exceeded. For irrigation and stock, a reliability of eight years out of ten is acceptable, and for domestic schemes the aim is nine years out of ten.

The selection of percentage yield within the given range depends on local experience. For example, the lower limits in the yield column usually include forests, areas of cultivation and improved pastures.

The run-off, in megalitres, from the catchment is calculated from Table 1.2 and according to the following formula:

$$V = K \times A \times R \times Y$$

where: V = run-off yield (ML)
   K = conversion to megalitres = 0.01
   A = catchment area (hectares),
   R = average annual rainfall (millimetres), and
   Y = yield (using Table 1.2, and expressed as a decimal
      for example, 12.5% = 0.125; 7.5% = 0.075).

Table 1.2 Yield from natural catchments.

Average Annual Rainfall (R) (mm)	Total Annual Evaporation (mm)	Yield as percentage of average annual rainfall (Y)				
		Reliability (years out of ten)	Shallow sand or loam soils (%)	Sandy clay (%)	Elastic clay (%)	Clay pans or inelastic clays (%)
>1100	<1000	8	10 to 15	12 to 20	15 to 25	15 to 30
900 to 1100	<1000	8	10 to 15	10 to 15	12.5 to 20	15 to 25
500 to 899	1000–1300	8	5 to 8	7.5 to 12	7.5 to 15	10 to 15
500 to 899	1300–1800	8	5 to 8	5 to 12	5 to 12	10 to 15
400 to 499	1300–1800	8	3 to 4	3 to 7.5	4 to 6	7.5 to 12.5
250 to 399	<1800	8	1.5 to 3	1.5 to 5	1 to 3	2 to 5
<250	>1800	8	1 to 2	1.5 to 3	1 to 2	2 to 5

Source: Modified from Burton, 1965 and Nelson, 1983.
Notes: i For 9 out of 10 years reliability multiply percentages by 2/3.
　　　ii Numbers should be halved if catchment sown to improved perennial pastures.

Example

A small catchment of 50 hectares is forested and the soil is a sandy clay. It receives an average annual rainfall of 1000 mm and has an evaporation of 1000 mm. What would the estimated run-off yield be for a domestic and stock scheme?

$$K = 0.01$$
$$A = 50\,\text{ha}$$
$$R = 1000\,\text{mm}$$
$$Y = 10\%, \text{ or } 0.10 \text{ (reliability 8)}$$
$$V = \text{run-off yield (ML)}$$
$$\text{Run-off yield } V = 0.01 \times 50 \times 1000 \times 0.10$$
$$= 50\,\text{ML}$$

For normal stock dams, and for the irrigation of annual pasture, an 80 per cent reliability is generally used. An 80 per cent reliability yield means that the catchment yield or run-off will be exceeded 80 per cent of the time, for example, in four out of five years. If the water is for domestic use, or is to be an adequate drought storage, a 90 per cent reliable yield should be used. Alternatively, additional water should be stored to provide water during the critical storage period, that is, the period of no run-off. However, in the absence of reliable data, the 'rule-of-thumb' method can be applied to estimate the reliable yield, which is defined as the percentage of the total volume of water that falls on the catchment during the year. While it is acknowledged that this method is basic and arbitrary, it is also apparent that it will continue to be used in small rural catchments. This estimate is of critical importance when landholders request a storage of optimum capacity. In this method, the percentage yield is calculated using the factors of average rainfall and soil type, derived from Table 1.2. In those parts of southern Victoria which are subject to prevailing westerly influences in the winter months, the annual pattern of run-off, while not so sharply defined as in the north,

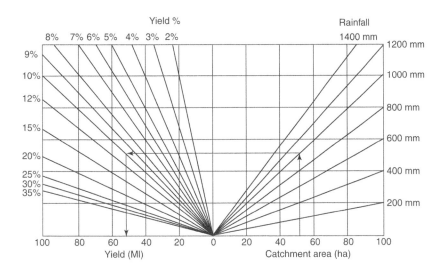

Figure 1.1 Water Yield (*Source*: Beavis and Lewis, 1999).

involves a winter season. This is when replenishment of farm storages can normally be expected, followed by a comparatively dry summer and autumn.

It should be noted that estimations of yield from these tables are largely tentative. They may be over-conservative in some districts and may be modified if good evidence of their unreliability exists. Any increase in percentage run-off should, however, be accepted with caution, bearing in mind that the values in Table 1.2 are for 80 per cent reliability conditions and are not mean values. Their usefulness is that they represent a 'first-cut' methodology in assessing water resources in the field.

The estimation of yield (V) ML in the example can also be shown by using Figure 1.1 once the catchment area (A) hectares, annual rainfall (R) mm and yield as a percentage (%) in Table 1.2 has been selected. Simply place a mark on the catchment area (ha) axis, trace up to rainfall (R), then move across to yield % that has been obtained from Figure 1.1. Directly below is the storage volume or yield (V).

The run-off yield method can be used to take into account existing and future land use even if the land will be periodically rotated after cropping. The end results are not affected, particularly when irrigation areas are taken out and then used for stock and domestic purposes. The basic criteria used are fixed and based on catchment area (hectares), application rate (megalitres per hectare) and annual rainfall (millimetres). The only variable is the yield (percentage). This will vary only if the land has been subdivided for closer settlement and more intensive farming practices. Figure 1.1 can be used to determine the run-off yield for small catchments, given the storage capacity of an existing dam when it is full without overtopping.

The storage design procedure should be based on the provisions of sufficient storage capacity to meet water demands over a selected 'critical storage period'. Usually, in both case study areas (northern and southern Victoria) it is the period from October to March, inclusive. This is the period without appreciable inflow when the storage must supply all water demands and losses without replenishment. For design of minor storages, 'rule-of-thumb' values of critical storage period are useful. For the design of larger storages, the critical storage period must be determined from an analysis of past run-off estimates. In both cases it is necessary to calculate a quantity, and time pattern, of water use for the critical period in order to determine the storage volume needed to meet all demands and losses over this period.

1.2.3 How trees affect yield

The catchment for every small dam is unique. In a small catchment, or where rainfall is low, widespread tree cover will reduce yield. In a large catchment, or where rainfall is high, water yield is usually high and the amount of tree cover is less relevant. In most cases, a tree cover of approximately 5–10 per cent comprising shelterbelts, small woodlots or scattered trees is unlikely to have any noticeable effect on water yield. If water yield from a small dam catchment covered in good pasture were compared with yield from the same catchment with 5–10 per cent of mature tree cover added, the difference in water yield would be hard to measure. At the same time, the trees provide many other benefits to water quality, pasture and crop production (Greening Australia, 1990).

1.2.4 Artificial catchments

Relatively low percentage yields have made engineers attempt to produce larger yields by treating catchments. In the United States these techniques are called 'water harvesting', in Australia, the preferred term is 'artificial catchments'.

There are past records of humans modifying the landscape to augment surface run-off for agricultural purposes. Ancient cultures in the Middle East and Europe had an extensive system of channels and aqueducts to collect run-off and transport the water to irrigation areas.

A practical technique of water harvesting has been developed in Western Australia. The catchment area is graded, compacted into a series of parallel roads, and then drained into a large channel leading to the storage. However, there are obvious issues of soil erosion using this technique. Technical analysis of dimensions and slopes is given in Hollick (1975).

In some regions, traditional construction materials such as sheet metal and concrete have been used to make artificial catchments. Gibraltar, for example, is totally independent of outside drinking water supplies and has 40 hectares of metal roofing, concrete paving and treated rock surface on the eastern slopes of the Peninsula. The run-off is collected by gutters and passed to dams.

Other materials include plastic and metal films that are either anchored or bonded to the soil, or cheap bitumen paving. Unfortunately, oxidation of the bitumen frequently means that the run-off is discoloured, but work on this material still continues. Chemical treatments to reduce the permeability of soil in a catchment are another means of increasing yield. However, they do not last for long and usually increase the erodability of the soil.

Construction costs vary from site to site for different treatments:

- a roaded catchment could cost as little as $500/hectare where there is clay close to the surface, or as much as $3000/hectare if there is a lot of sand and gravel;
- a bitumen treatment could range from $500/hectare to $25,000/hectare and last for five years to ten years respectively.

In common with other construction works, the unit cost will decrease as the area of treated surface increases.

1.3 DAM SITE SELECTION

Surveys have indicated that a large proportion of small dams fail because of poor planning, unsatisfactory siting, faulty construction or lack of maintenance. Such costly failures can usually be avoided. The choice of a suitable dam site should begin with preliminary studies of possible sites. Where more than one site is available, each should be studied separately with a view of selecting the one that proves most practical and economical (DoA Vic, 1978; SRW, 1995).

From an economic viewpoint, a dam should be located where the largest storage volume can be obtained with the least amount of earthworks. This condition will occur, generally, at a site where the valley is narrow, side slopes are relatively steep, and the slope of the valley floor will permit a large deep basin. Such sites tend to minimise the area of shallow water. Except where the dam is to be used for wildlife, large areas of shallow water should be avoided due to the potential for excessive evaporation losses. Value of the land flooded by the storage should also be considered.

Dams to be used for watering livestock should be spaced so that livestock will not have to travel more than 200 m to reach them in rough, broken country, nor more than 1,000 m in smooth, relatively level areas; that is, the maximum spacing between dams is 1 km. Forcing livestock to travel long distances for water is detrimental to both the livestock and the grazing area. Overgrazing near water, and unused feed far from water, are characteristics of inadequate water distribution.

Where water must be conveyed for use elsewhere, such as for irrigation, fire protection or stock and domestic use, dams should be located as close to the point of use as practical. The economics of gravity flow compared with pumping must also be considered.

Pollution of small dam water should be avoided by selecting a site where drainage from houses, piggeries, dairies, sewerage lines and similar areas will not reach the dam. Where this cannot be done practically, the drainage from such areas should be diverted from the dam.

The dam should not be located where sudden release of water due to failure of the dam would result in loss of life, injury to persons or livestock, damage to residences, industrial buildings, railroads or highways, or interruption of the use or service of public utilities. Where the only suitable site presents one or more of these hazards, a more detailed investigation should be made.

Powerlines present a hazard to people constructing, using or desilting small dams. Sites under such lines should be avoided. Permission from the electric supply company is recommended before construction is commenced beneath powerlines.

1.3.1   Choosing a dam site

When choosing a dam site (Lewis, 1995b), the following points need to be considered:

i Storage yield from the catchment

Yield is the volume of water harvested from the dam catchment area. It depends on rainfall, plant cover, slope, soil type, area and other factors (see Section 1.2.1).

Three questions need to be asked when selecting a dam site:

1   what is the catchment area above the dam?
2   how much water will the catchment yield?
3   would the catchment yield be substantially reduced if another dam were to be built in the same catchment?

If most of the catchment is outside the property and is eroded, a silt trap may have to be built. This is usually a small dam above the main storage (Figure 1.6). Its function is to slow down the run-off so that silt is deposited before it is carried into the main dam. Silt traps have to be cleaned out periodically.

ii Increased catchment harvest

Often dams cannot fill because the catchment area produces insufficient run-off. The catchment, or source of water of a dam, should generate enough water each year to fill it.

If the catchment is large enough, graded drains can be constructed to divert run-off from adjacent areas. Hard surface areas such as roads and roofs can also be used to increase the yield (see Section 1.2.4).

iii Choosing dam type for site

The larger the quantity of water stored for each cubic metre of soil to be moved, the larger is the cost-saving on construction. Wherever possible, a site with a good storage excavation ratio (S:E) should be selected. The average stock dam site would have a storage to excavation ratio of 2 or 3 to 1, that is, 1 m^3 of soil moved for each 2 to 3 m^3 of water stored (see Section 1.4).

iv Soil assessment and testing

Although clay soils are imperative to prevent leaking dams, it should be noted that all clays do not hold water. Physical and chemical properties can make some clay soils prone to seepage or tunnelling, both of which result in bank failure.

To test the suitability of the soil, samples should be taken from the borrow pit (see Section 2.1.2) and along the bank centre line using a soil auger, which penetrates at least just below the proposed excavation depth. These tests show whether or not the soil contains clay. Further tests are needed to find if this clay is of a water-holding type.

If testing indicates that the soil is suitable in terms of clay content and type, the following points should be checked:

- Are there any sand or gravel layers or seams? A dam should not be sited where there are seams, fissured rock or soaks. A stream or soak is an indication that the water-holding layer has not been reached. Where this occurs, it will be necessary to dig down to the impermeable layer and repack with a suitable clay to stop water leaking out along the original soak line.

- Will the clay content of different soil samples be sufficient to seal the bank against leaks? Soils from different locations should not be mixed to form a composite. Test results from such a sample can be meaningless because even though the mixture may hold water, one of the soils in the mixture may leak. The dam will then leak where this soil is exposed in the excavation.
- Is there sufficient material with correct clay content readily available within the excavation area? Unless extra care and time are spent, there will almost certainly be layers of unsuitable material in a borrow pit which will leak when placed in an embankment. It is better to keep the soils separately zoned, as explained in Section 3.1.1.
- Is the layer of permeable topsoil overlying the clay soil too deep? If there is more than 1 m of topsoil to be removed, costs should be estimated carefully. Removal of topsoil may become a large proportion of the total costs.
- If the dam is to be filled with groundwater, special care must be taken in testing soils. Some groundwater contains chemicals which react with clays and cause leakage even though the soil is watertight to rainwater.

If you have any doubts, seek advice from an experienced soil/small dams engineer.

v Planning outlet structures

Many dams fail because of an inadequate or incorrectly located spillway or insufficient freeboard. If a spillway is too small to cope with storm flood-flows, water will flow over the top of the bank which may then breach. A badly designed or constructed spillway can cause erosion of the spillway and lead to complete failure of the dam. Therefore, the spillway should be large enough to handle and dispose of flood-waters safely, without damaging the dam bank or causing erosion of the spillway.

When planning to divert water a landholder should make sure it does not leave the property in a different drainage line from its natural alignment. If it is advantageous to do this then it would be wise to obtain permission from down-slope neighbours and, if road works will be affected Ñ from the municipality. The permission for this diversion should be in writing from all parties concerned to prevent future arguments.

In some cases, flows from the catchment are so great that there is no economical way of constructing an in-stream dam with an adequate spillway. In these cases consideration should be given to an off-stream storage. Where it is anticipated that water may flow over the spillway for a prolonged period, for example one week, problems may occur due to deterioration of the vegetation cover on the spillway. This problem can sometimes be overcome with the installation of a trickle pipe to carry the prolonged flow.

vi Legal considerations

Before constructing any dam across a gully or depression, it should be checked first that it is not a legally recognised waterway, and that it does not have an existing easement. This may be found on the land title or in the records of the local municipality. If the gully or depression does have either or both of these conditions, then permission must be obtained from each responsible authority, agency, instrumentality or department. Permission should also be obtained from any neighbours who may be affected.

In some States of Australia, the construction of a dam on a river, stream, water-course or waterway, is governed by common law and/or legislation. In addition, a separate licence may be required to take and use the water that is to be stored in the dam.

For details on owning a dam as an asset or a liability see Section 10.

vii Environmental issues

It is preferable to design and build in-stream dams in a manner that inherently provides for fish passage. However, where existing dams are to be modified, or where it is not practical to modify the design of new structures to provide for fish movement, it will be more appropriate to install some type of dedicated fish way (Lewis and O'Brien, 2001). This decision will be based on a number of site-specific factors such as the fish species present, site topography, flow characteristics and the potentially increased costs. Over the years, a large number of in-stream structures have been built without any provision for fish movement. Some of these are large dams and weirs that require a fully engineered device such as a fish lift or a vertical-slot fish way to facilitate fish movement. However, the vast majority of barriers are smaller structures for which other options, such as rock fish ways, are considered more appropriate. There is also a need for those planning to install an in-stream structure to conform with the current legislative requirements and practices (Lewis *et al.*, 1999 and O'Brien *et al.*, 1999).

Conditions regarding installations of in-stream structures have been formulated over many years in Victoria to minimise the impact on aquatic biota and maintain their natural environment. This Section does not discuss the legislative responsibilities associated with providing fish passage, but does demonstrate graphically an in-stream structure that assists fish passage (see photo 1.1).

viii Improving biodiversity and aesthetic values

Finishing touches to a dam include:

- fencing the entire pond area to keep livestock out, and installing drinking troughs away from the dam. This minimises pollution entering the storage.
- planting trees and shrubs to provide windbreaks which prevent wave action and therefore soil erosion, while also providing shelter for wildlife (do not plant trees on the banks as it will create seepage through the banks).
- stocking the dam with suitable fish which can provide food and recreation.

1.4 TYPES OF FARM STORAGES

Small dams are small earth structures designed to impound water for stock watering, domestic supply, aquaculture, irrigation or as a component of an integrated erosion control network on agricultural or pastoral land.

Photo 1.1 Typical in-stream barrier with rock-ramp fish way, Merri River, Victoria (Photo: T. O'Brien).

Dams can be divided into two main types:

1   Drainage line or unregulated waterway dams include those dams located in drainage lines or waterways which are filled by run-off from the catchment upstream;
2   Off-waterway dams include structures which are filled by a diversion channel or by pumping from a waterway or groundwater.

A stock, domestic or irrigation dam can be of either type. The purpose of the dam influences the shape that is required and this in turn influences the kind of site that is best. For instance, a stock dam must be deep enough to allow for evaporation and still have enough water left over for the stock. Evaporation losses from a shallow storage are important and can easily amount to 30 per cent over a hot summer (see Section 1.5.1). In a deep storage these losses may be only 15 per cent. Therefore, the ideal site for a stock dam is where there is adequate depth of a suitable soil. The same applies for fire fighting and domestic dams.

Where security of supply is important, two years supply of water should be stored to guard against a drought year. In most parts of Australia loss to evaporation over the two years can be minimised by a greater depth of storage. The ideal site for this kind of dam is one with a good depth of suitable soil as close as possible to where the water is to be used.

The ratio of the volume of water which can be held to the volume of the earth excavated (m^3) to form a small dam, is one means used to compare different storages and storage sites. This ratio is called the storage : excavation ratio (S:E). High ratio values imply a low capital cost of storage. Many landowners seem to have the wrong idea when looking for large storages. A common belief is that a short, high embankment across a deep valley will store many megalitres. This may be true, but a high bank is costly to build. Usually a long bank, which is low in comparison to its length, will store more water for less capital outlay. The reason for this is that when the height of bank is doubled, the cost of the dam increases four times. However, when the length is doubled, the cost of the dam is doubled. This last point may be disputed since evaporation losses will be very high with this type of storage. The objection is real, but the fact is that deep, narrow valleys with a flat floor are unusual. Furthermore, a narrow valley will not hold as much water as can be stored over a wider area. Obviously, a one metre depth of water over a one hectare surface area, in a narrow valley cannot possibly be equal to a one metre depth over five hectares on a flat site.

There are many types of farm storages constructed, but the terminology in describing them is not standardised. A selection with associated design and construction details are listed in the following sections.

1.4.1   Gully dams

A gully dam consists of an earth embankment built across a waterway, valley, depression or drainage line (Figure 1.2). The embankment generally incorporates an earth spillway at one or both ends to pass surplus water. The earth spillway is sometimes supplemented by a pipe spillway (trickle pipe), which helps protect the earth spillway from the effects of long-term flows.

Outlet pipes (also called compensation or low level pipes) are commonly provided through the embankment to service stock, domestic irrigation or environmental flows.

Conditions when most used

These dams are normally built from material located in the storage area upstream of the dam site. If possible, excavation should be above the level of the outlet from the reservoir to maximise yield. The ideal site for a gully dam is where the sides of a valley are close together and then widen out above the point where there is flat ground.

If the slope of the valley floor is sufficiently flat a relatively low dam will impound water over the natural surface, for a considerable distance upstream. An S:E ratio of up to 10:1 can be achieved at a favourable site.

Gully storages must incorporate some means of passing flood flows. They do, however, provide the most economical form of storage and are particularly suitable for irrigation development as well as for stock and domestic purposes.

1.4.2   Hillside dams

An earth dam located on a hillside or slope, and not in a defined depression or waterway, is called a hillside dam and is usually three-sided or curved (Figure 1.3).

Conditions when most used

To achieve the maximum S:E ratio the embankment should cross the natural contours at right angles. On flatter terrain the hillside storage can have an S:E ratio of up to 5:1 and is suitable for irrigation use in areas of low evaporation. On steeper slopes the S:E ratio can become less than 1:1 and long downhill batters are necessary to ensure

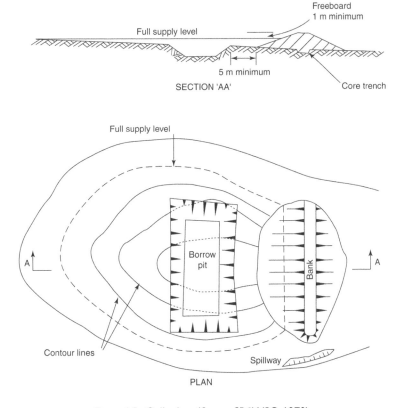

Figure 1.2 Gully dam (*Source*: SR&WSC, 1970).

embankment stability. Material for the bank is usually taken from within the storage area. As with gully storages, provision of a spillway clear of the actual embankment is needed to pass surplus flows from the natural catchment.

The inflows to hillside storages can often be increased by the construction of catch drains, or under favourable circumstances, by diversion of surplus stream flows in addition to natural run-off.

1.4.3 Ring tanks

A ring tank consists of a storage confined entirely within a continuous embankment which is built from material obtained within the storage basin (Figure 1.4). Consequently, water is held partly above and partly below the natural surface. This is achieved by radial bulldozing or a scraper working in annular ring borrow pits. Compaction equipment is also necessary. Ring tanks are filled by pumping from groundwater or streams during periods of surplus flow. The size is therefore determined by the pumping plant available and the periods of surplus flow.

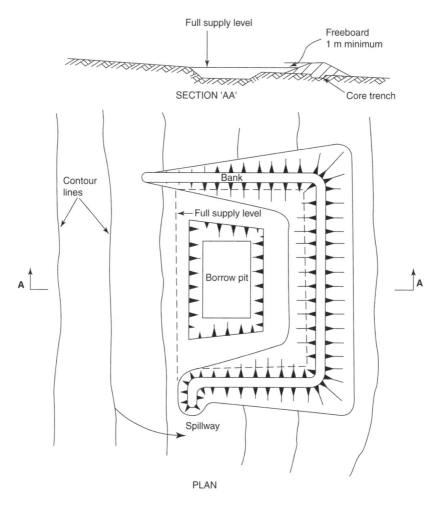

Figure 1.3 Hillside dam (*Source*: SR&WSC, 1970).

Conditions when most used

Generally the surface area is large and, in relation to depth and high evaporation rates, losses can be a problem if surplus stream flows are infrequent or irregular. The S:E ratio increases with diameter, ranging in practice from approximately 1.5:1 for a small tank to 4.5:1 for a large tank.

Ring tanks are used primarily for irrigation in flat terrain. They have the advantage that inflow and outflow can be controlled without expensive spillway provision. In pumping from streams, use can be made of catchments which might otherwise be unavailable.

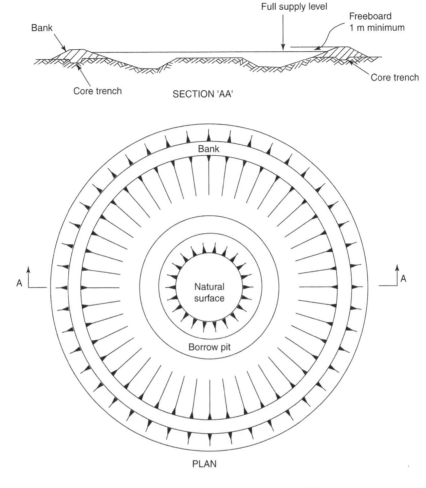

Figure 1.4 Ring tank (*Source*: SR&WSC, 1970).

1.4.4 Turkey's nest tanks

The ring tank is often incorrectly called the 'turkey's nest tank'. A true turkey's nest tank is constructed from material borrowed from outside the storage area (Figure 1.5). All water is therefore held above ground level and is made available by gravity through an outlet pipe in the lowest point of the embankment.

Conditions when most used

Although storage water can be discharged through a low level outlet pipe, pumping is still required to fill the storage. These storages are usually much smaller than ring tanks and have a lower S:E ratio. The principal use of turkey's nest dams is as a balancing reservoir between a pump supply, windmills, and stock troughs.

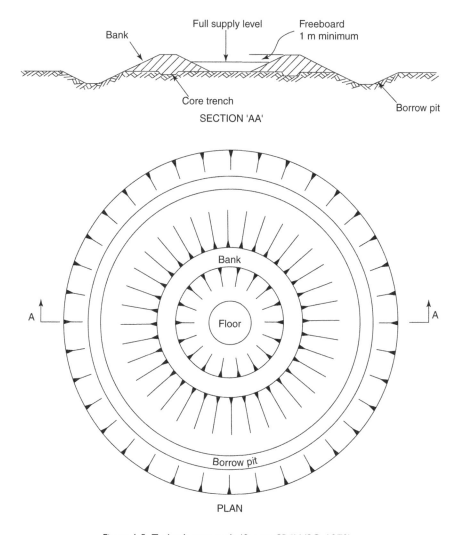

Figure 1.5 Turkey's nest tank (*Source:* SR&WSC, 1970).

1.4.5   Excavated tanks

Excavated tanks are restricted to flat sites and comprise excavations below the natural surface in locations where suitable sites for gully and hillside storages do not exist (Figure 1.6). The excavated material is wasted, and is usually placed adjacent to the excavation in neat spoil banks. Although excavated tanks can be any shape, they are usually in the form of an inverted pyramid.

Conditions when most used

An excavated tank that is filled on an annual basis only should be as deep as possible (3 m minimum) to reduce the surface area in relation to depth in order to reduce the

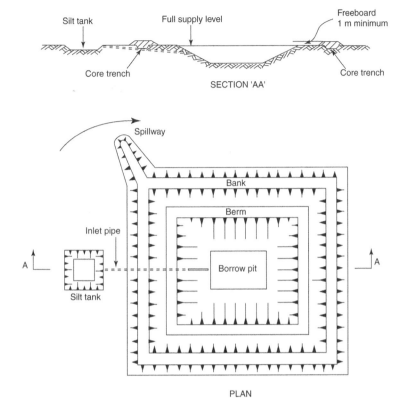

Figure 1.6 Excavated tank (*Source*: SR&WSC, 1970).

effects of high evaporation. Catch drains can be constructed to collect surface run-off which, on sloping country, can raise the ponded water above surface level.

When the storage is gravity filled by surface run-off which contains appreciable quantities of silt, it is desirable to divert this run-off into a small excavated silt tank about one-tenth the size of the main storage. This is to reduce the velocity of the incoming water so that the silt is deposited in the silt tank rather than in the main storage. A suitable pipe connects the two tanks.

A storage collecting water from its own catchment requires a spillway, or by-wash, to pass excess run-off. The catch drains are generally constructed to incorporate the spillway, which must be located so that the overflow does not damage the tank walls. A silt tank and spillway are not required for off-stream storages that are filled by pumping or a controlled artificial channel.

Excavation tanks are generally limited to stock and domestic uses, and the irrigation of high value crops, because of low S:E ratios and the high unit cost of water storage. The excavated storage may require fencing to prevent fouling and pollution by stock. Troughs supplied by windmill or pump are then provided for stock watering.

1.4.6 Weirs

A weir is a structure which is not used to store water but to raise the upstream water level to allow diversion into off-waterway storages or create a pump sump. It can be constructed of earth, concrete or timber.

Conditions when most used

Small weirs are quite frequently used as farm storages, usually where most of the water supply required is available from low flow in the waterway concerned. In the past, these have all been hard weirs, built to resist the erosion effects of high velocities and high-energy losses over the weir. Because of the expensive construction materials, these weirs are located at sites with a short crest length, where a rock bar or other feature causes a natural constriction of the stream.

A valid criticism of weirs on waterways with a low flow supply is that they may be used to obtain an excessive share of the flow. To avoid this, all weirs require a bed outlet, so that once the storage is used, remaining low flows can be passed to downstream users. Operation of this is difficult to police, resulting in private weirs generally not being favoured in major streams. However, if operated correctly, a weir can provide useful storage to supply irrigation or stock water during periods of insufficient low flow.

1.4.7 Off-waterway storages

Off-waterway storages are used to store water that is diverted or pumped from ground-water, an adjacent waterway or catchment. These off-waterway storages are usually in the form of a gully dam, hillside dam, ring tank or excavated tank. They usually have either very little direct catchment, or none at all.

Conditions when most used

An off-waterway storage has the advantage of having fewer foundation problems and, if the waterway has a large catchment, that it does not need an expensive spillway.

1.5 DAM STORAGE SIZE

In order to determine the required capacity of a dam, it is necessary to allow for environmental requirements, and losses such as evaporation and seepage. Accurate calculations of the losses can be complex. However, as an approximation, it is advisable to allow for losses in the order of 40 per cent of total storage volume, depending on locality and weather conditions. This means that only 60 per cent of water stored is available for consumption, and this is commonly referred to as the useable volume (Figure 1.7).

During low rainfall periods, replenishment rates can be too low to replace consumed water. Therefore, storages need to be large enough to accommodate these periods. Dams should be able to cope with drought periods at least equal to 18 months, including two summer seasons (see Section 1.2.2).

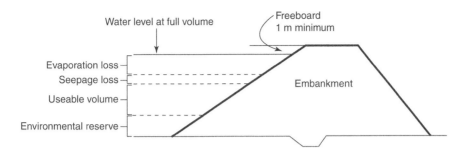

Figure 1.7 Water losses and needs.

When a pump or windmill must be used to supply dam water, it is advisable to:

- make sure the storage is large enough to meet anticipated daily needs; and
- locate the dam in a position that allows supply to the whole farm over two days by gravity feed, even if at reduced water pressure.

1.5.1   Evaporation losses

Evaporation can significantly reduce production. More than 40 per cent of water can be lost by evaporation from stored water in a 12-month period in certain locations. This loss will reduce the amount of water available for irrigation, with possible significant production losses.

Since evaporation is often the biggest consumer of water from a dam, it must be taken into account when choosing dam size. Evaporation will vary according to climatic zone, time of year, dam size, dam shape and the specific location of the dam.

A first approximation of annual loss to evaporation can be calculated from the following relationship:

$$L_E = 0.67E \times A_F$$

where L_E = evaporation loss (L)
E = local annual evaporation (mm)
A_F = surface area of the dam at full supply level (m^2)

To take an example: for a dam with a surface area of 5000 m^2 (0.5 hectare) and an annual evaporation of 1275 mm, the volume of water lost from the storage through evaporation is estimated as:

$$L_E = 0.67 \times 1275 \times 5000 = 4\,270\,000\,L \text{ or } 4.27\,ML$$

Evaporation, however, varies considerably through the year. During the summer it is usually about twice that of either the spring or autumn months. Hence the three summer months account for approximately half of the yearly total. Estimation of evaporation for periods of less than a full year needs to take this seasonal variation into account.

The Bureau of Meteorology in conjunction with CRC Catchment Hydrology, on 5 July 2001 released two books, titled *Climatic Atlas of Australia – Evapotranspiration* and *Climatic Atlas of Australia – Rainfall*. This information could be used to assess both evapotranspiration losses and rainfall.

1.5.2   Ways of controlling evaporation

Various means of controlling evaporation have been suggested, including chemical films, floating or suspended material, windbreaks and emergent water plants. However, more information is needed in order to measure the relative efficiency of these methods.

A recent study by RMIT University, Victoria, into aspects of evaporation from off-waterway storage, recommended the following ways of decreasing evaporation:

i   Floating water plants do not achieve significant savings in evaporative losses.
ii   Floating rings with reflective caps significantly reduced evaporative losses, and further development of this option is proceeding. The rings are formed from strips of buoyant plastic, covered with a cap of bubble plastic and painted with white reflective paint.
iii   Initial tests indicated that windbreaks have potential for reducing evaporation, but further development of this idea is needed.
iv   Two simple methods for reducing losses are to have deeper storages, and to split large storages into cells.

1.5.3   Seepage losses

Pervious sub-surface soils are not always detected in the initial investigation because landowners cannot afford, or do not understand, the necessity for comprehensive soil testing of the base of a storage.

The borrow pit is usually located beneath the proposed storage. Material from this source that is used in the embankment is not necessarily impervious and may result in seepage.

When the borrow pit is within a storage there are underlying strata that may not be watertight. Excavating these pervious seams will result in seepage and leakage through the base. One problem that often occurs with high storage ratio dams, in wide flat bottomed valleys, is that they may contain a considerable depth of sand, gravel or silt.

For a simple field method of testing for seepage in the borrow pit and other areas, see Section 2, and for remedies Section 6.

1.5.4   Average water consumption

The information provided in the compiled figures below (Tables 1.3 to 1.7, Figures 1.9 and 1.10), is based on average conditions using Department of Natural Resources and Environment (Victoria) data and Agnotes (see also Table 7.2 and SRW, 1995 – Small Dam Series of pamphlets). In most cases the data can be applied to other areas, by adjusting for local climate and other variables.

Table 1.3 Annual rainfall and roof run-off relationships.

Rainfall (mm)	Roof area (m²)			
	150	200	250	300
	Volume (litres)			
100	15 000	20 000	25 000	30 000
400	60 000	80 000	100 000	120 000
600	90 000	120 000	150 000	180 000
800	120 000	160 000	200 000	240 000
1000	150 000	200 000	250 000	300 000

To achieve 100 per cent run-off, the following need to be considered:
• spouts and downpipes have sufficient capacity to deal with peak flows, otherwise losses occur;
• small evaporation losses do not occur from the roof.
(Information based on various roof areas and a range of climatic conditions.)

Table 1.4 Method of calculating dam or tank size (litres).

Month	Storage at start of month (L)	Inflow (L)	Outflow (L)	Difference between inflow and outflow (L)	Storage at end of month (L)
October	45 000	15 000	14 000	+1 000	46 000
November	45 000	12 000	14 000	−2 000	43 000
December	43 000	5 000	16 000	−11 000	32 000
January	32 000	5 000	16 000	−11 000	21 000
February	21 000	3 000	14 000	−11 000	10 000
March	10 000	10 000	13 000	−3 000	7 000
April	7 000	12 000	9 000	+3 000	10 000

Notes:
i In first month 1000 litres is lost as overflow.
ii It is necessary to work figures over two summer periods to establish number and size of storage(s).
iii On small farms (for example, 10 hectare) in areas with reasonable rainfall, large concrete tanks (45 000 litres capacity) filled from roof run-off are often adequate, on their own, to provide both stock and domestic water.

Table 1.5 Daily domestic water usage.

Type of use	Average daily requirement (Litres)
Household	180 per person
Garden	3,000 per 0.1 hectare

Note: based on summer months only.

Peak consumption rates tend to govern the size of dam storages, but it is difficult to estimate peak demands for stock. However, it can be assumed that average daily consumption may occur over 20 per cent of the day (approximately five hours). In domestic consumption the period is less than 5 per cent of a day (approximately one hour). Sizes based on these quantities are normally adequate for most installations.

Table 1.5 provides an estimate of how much drinking water stock will need in a week.

Table 1.6 Water application rates for different uses.

Culture	Application rate – average annual quantity of water required (megalitres/hectare)
Annual crops	3.0
Berries	3.0
Cereals	3.0
Citrus	3.0
Flowers	3.0
Golf Course	3.0 (south of Divide)
	6.0 (north of Divide)
Nursery	3.0
Lucerne	6.0
Market Garden	3.0
Orchard	3.0
Native Pasture	3.0
Tobacco	3.0
Vines	3.0
Permanent Pasture	6.0
Annual Pasture	3.0

Table 1.7 Daily animal water usage.

Type of stock	Average daily requirements (litres/day)
Sheep	7
Cows (milking)	70
Horses	50
Pigs	25
Poultry	32 per 100

Note: Does not take into account whether or not stock are on dry pasture (see also Table 7.2).

Table 1.7 provides estimates only, which are affected by:

- pregnant and lactating animals. These animals require larger volumes of water than non-brood animals.
- the age of the stock. Older animals need greater amounts of water than young or newborn animals.
- type of pasture and/or feed. Hand-feeding stock with dry feed increases their water requirements significantly.
- weather conditions (hot/cold/windy etc.).
- presence of shade (trees, shelterbelts, etc.). Shade and protection from the elements can help reduce stock water requirements.
- distance to water/availability.
- quality of water. Saline water will increase water consumption.

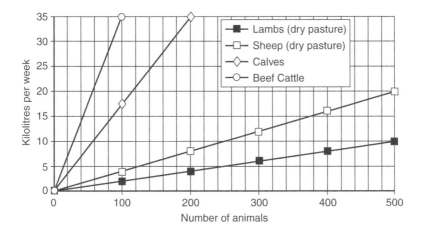

Figure 1.8 Assessing weekly needs for different livestock (KEY: 1 kilolitre = 1000 litres (or 220 gallons)).

1.6 USING A DAM IN DROUGHT

Every State of mainland Australia is subject to periodic droughts, which makes it necessary to cart water for stock and domestic purposes. A well-planned water supply scheme can alleviate such hardships. Below are some facts and some simple tips and techniques to help plan stock water use and water conservation during a long dry season.

Drought is a word that strikes fear into the hearts of dryland farmers across Australia. Even in so-called 'good' years, you may be asking yourself, 'Is there going to be enough drinking water for my stock?' and 'How can I manage my water in case of drought?'

To plan for drought effectively the following questions need to be addressed:

i   How much water does my stock need?
ii How long will my water supplies last?

Once these two questions have been answered, it will be possible to make informed choices about water and stock management. The first question can be answered by referring to Section 1.5.4, whilst some rules-of-thumb approaches to help answer the second question are given below (Figure 1.9)

The graphs in Figure 1.9 provide rough estimates of how long dam water lasts for dams with a maximum depth of 1.0 m, 1.5 m, 2.0 m and 3.0 m, taking into account evaporation losses. Note that if a dam is extremely silty it will reduce the volume of water by as much as 20 per cent.

To use the graphs:

•   measure the surface area of the water in the dam at the beginning of January;
•   choose the graph that shows the maximum depth for the dam being assessed;

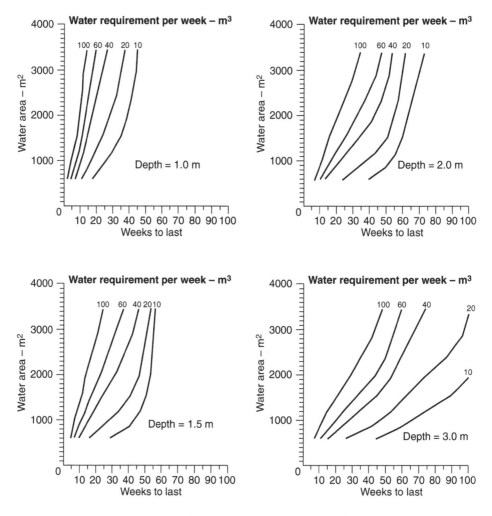

Figure 1.9 How long dam water lasts (*Source*: SRW, 1995).

• measure across the graph horizontally to the amount of your weekly water requirements, then draw a line to the bottom axis.

This will show the number of weeks you will have water in the dam, if no further rain falls during that time.

Example
Maximum depth January 1: 2 m
Surface area January 1: 1200 m³
Weekly water requirement: 40 m³

In this case the available water would last about 25 weeks, or until the end of June.

KEY:

kL = kilolitre
L = litre
m^3 = cubic metre
m^2 = square metre
mm = millimetre
m = metre
km = kilometre

If the graph shows that water supplies will be depleted before the next expected rainfall, a number of choices are available:

A Reduce/conserve your water usage

As a first step, this is probably the most cost-effective and easiest option to achieve. Below are some hints for making your water go further and to help manage your water resources during dry times:

- fence off springs, soaks and dams, and pipe water to covered troughs. This helps prevent losses by evaporation and bogging by stock as they go in search of water.
- pump water from shallow dams to one central dam to reduce evaporation losses.
- build drains from road catchments and other hard surfaces to dams, to catch any run-off from summer thunderstorms.
- de-sludge, deepen and/or carry out maintenance on empty dams while you have the opportunity, and to prepare for rain when it comes again.
- plant or build shelterbelts to reduce stock water losses through perspiration.
- use pumps and pipes that best suit your needs, to reduce wastage and pumping costs.
- if you use streams or bores for stock drinking water, check them for salinity levels. If they are saline, clean your troughs and uncovered tanks each week as they will corrode quickly.

B Cart water in

This is not only an expensive option, but can also be soul-destroying and time-consuming. When looking at carting water in, consider the costs.

- Costs of carting water vary depending on whether you hire a contractor or cart it with your own truck.
- Contract rates vary from $45 for 10 000 L (in areas where milk tankers are used for back-loading to dairy farms) to $60 for 4,500 L in areas where there are fewer contractors or supplies are harder to obtain.

When deciding whether to cart your own water, you will need to consider operating and labour costs (see example below).

Estimating costs of carting water

Example

If water has to be carted 9 km for 500 sheep, the daily costs could be calculated as follows:

Daily water requirement (from Tables 1.5 and 1.7, Section 1.5.4):

500 sheep × 7 L/day = 3500 L/day (770 gallons/day)

Truck operating costs:	18 km at $0.80 per km =	$14.60
Labour costs:	2 hours @ $15.00 per hour =	$30.00
	Total cost =	$44.60

This represents approximately 9 cents per sheep per day or $3 per sheep per month.

If you decide to cart water, there are some things you can do to optimise the use of that precious resource:

- store the water in a tank rather than a dam, if possible, to reduce losses by seepage and evaporation.
- round-shaped, above-ground swimming pools are the cheapest form of temporary storage for carted water;
- consider bulk cartage of water by a contractor – it could be much cheaper in the long run.

C Sell stock

This is a difficult decision but one worth planning well in advance, before stock start suffering and the decision is taken out of your hands. If you are unsure where to start, contact the nearest Agriculture Department for advice.

1.7 FIRE FIGHTING

Rural water and country fire authorities suggest an approximate supply of 15 000 L (3300 gallons) for fire fighting at a rural house. This represents about one-third of the capacity of many rainwater tanks on rural properties. This quantity is considered to be additional to other water uses such as domestic supply. This is important because the highest fire danger periods often occur during drought when water is scarce.

A supply of 15 000 L is considered a minimum quantity, particularly if buildings are located in bushland with a high fire risk. At least 50 000 L may be needed to protect a house in an isolated location.

Rural water authorities suggest that if household tanks are used, two outlets should be provided – an upper outlet for domestic use and a lower one for fire fighting. If the tank has a capacity of 45 000 L, the outlet for domestic use would need to be about one-third the height of the tank from its base. Unfortunately, rainwater tanks are not manufactured with this feature and in practice, additional water for domestic use often has to be carted in during drought periods.

If a dam is used to provide water for fire fighting, it may need to have a 50-year reliability, that is 2 per cent chance of such an event occurring in a 50-year period (the figure often used for protection of houses from flood damage). This is because of the strong temptation to use all of the water during drought periods, particularly if water for livestock or domestic use is scarce. This was found to be the case during the 1983 bushfires in Victoria, when dams had to be filled by water freighted in by rail.

Maximum daily consumption

Information from fire fighting authorities suggests that volumes of water required for bushfire control should range between 12–40 L/m^2 for buildings, and 7–25 L/m^2 for grass areas. This would typically mean about 14 400 48 000 L for a homestead and shed and 70 000–250 000 L for 1.0 ha of grass. The volume of water needed to control forest fires would be much greater. These quantities should be considered as the minimum amounts used within a 24-hour period.

Peak rate of demand

A minimum flow rate of 155 to 180 L per minute at a pressure of 240 to 340 kilopascals is suggested in most references, depending on the fire risk. In steep forested areas it may be desirable to reticulate water supply from a single large dam at greater pressures and flow rates to several properties. The need for a reticulated town water supply was demonstrated during the 1983 fires in Victoria and the 1994 fires in New South Wales.

Fire fighting literature states that up to 340 L per minute for a hose is needed for fighting forest fires. It is important to note that many small domestic pumps and garden hoses connected to a gravity tank may have adequate pressures but insufficient flow rates for fire control.

Country Fire Authority advice should be sought on the above points.

Monthly and annual consumption

One method of reducing fire risk is to burn off grass surrounding a homestead. This burn-off may need 40 000 to 250 000 L per hectare of grass burnt each year, depending on the difficulty of controlling the burn.

It is recommended that at least 15 000 L be provided each year for protecting a house in accessible locations and at least 50 000 L in isolated locations. If a dam is to be used to provide water for fire control in bushland, at least 500 000 L is suggested.

1.8   SMALL DAMS AND TREES

A Small dam is a long-term investment and proper design and construction will pay future dividends. Strategies to lower farm costs and improve water yield and quality include the planting of appropriate trees, shrubs and grasses in the right places around small dams to provide protection from wind and sunshine.

Dams are primarily intended to provide water for stock, irrigation or domestic use and often provide greater flexibility in pasture and stock management. Important secondary uses can include soil erosion control, bushfire protection, wildlife habitat, recreation and aesthetics. Whatever the purpose of the shelterbelt zone, the inclusion of trees, shrubs, grasses and other groundcovers in land and water management plans can provide the following benefits:

- a filter for cleaning water
- shade and shelter to reduce evaporation
- shade and shelter for stock
- emergency fodder and timber
- habitat for wildlife
- soil erosion control
- salinity control, and
- improved farm appearance (Greening Australia, 1990).

1.9 DAM COST JUSTIFICATION

The money spent on storing water in dams varies according to how the water is to be used (see Section 3.6). A suitably sized stock dam may make the difference between carrying 1000 head of sheep, or none at all. Similarly, a fire fighting dam may help to prevent thousands of dollars of damage. In these cases an S:E ratio even as low as 1:1 is justified (DoA SA, 1967). For irrigation however, more consideration is warranted. Again, the capital outlay for a certain amount of water varies with the crop that is to be grown. A vegetable or fruit grower can afford more expensive water than the person who will use it to grow pasture. In every case the landowner should consider whether the money would be better spent on pasture improvement or fencing. In general, irrigation should not be undertaken until a whole farm plan has been developed.

For permanent pasture irrigation from a catchment dam, a rule-of-thumb method for assessing storage cost is that the maximum economic outlay will be approximately $2500 per 5 ML. Therefore, at normal earthmoving costs this means an S:E ratio of 6:1. In the case of horticulture it pays to store water at an S:E ratio as low as 2:1, while stud breeders should consider an S:E ratio of 4:1 approximately. Below this ratio the costs should be worked out very carefully, as the dam is only part of the outlay. Distributing the water can entail heavy additional expense. Irrigation for drought insurance alone is unlikely to pay on any ratio.

When underground water is available, the cost of building a dam can be compared with the costs of sinking a bore. If there is good quality water underground in reasonable quantity, it takes a very favourable irrigation dam to outclass a bore. It is the high evaporation loss that makes dam water expensive. The unreliability of run-off to fill a dam may also favour a bore. In the case of dams filled by pumping, filling the dam adds further to the cost of stored water.

It is difficult for the average landowner to predict whether or not a water conservation and irrigation project will pay. This is because some landowners lack detailed facts and figures on the likely increase in production to be expected from a change in irrigation method (DoA SA, 1970).

However, a good method for comparing irrigation with other investments is as follows:

i   calculate the capital value and net returns of the property in its present state.
ii   work out the net returns as a percentage return on capital.

iii determine the extra capital involved in setting up the irrigation project, and add this to the present capital.

iv estimate the increased returns due to irrigation, not forgetting to deduct pumping costs, depreciation, extra labour, fertiliser costs and loss of production from land flooded by the dam. Add this figure to the present returns from the property.

v calculate a new percentage return, using the new capital and return figures.

Steps iii, iv, and v are then repeated for the other proposed investment or improvement, and this new percentage return on capital is worked out. Whichever project gives the higher percentage return is the better one from an economic point of view.

Investigation

The suitability of a dam site depends not only on the embankment but also on the ability of the soils in the storage area to hold water. The soil profile should contain a layer of material that is sufficiently impervious and thick enough to prevent seepage losses. Excellent materials for this purpose are clays and silty clays, although sandy clays are usually satisfactory.

Coarse textured sands and sand-gravel mixtures are highly pervious and therefore generally unsuitable. In some areas, such as flood plains, it is often possible to impound a limited depth of water in areas where no impervious layer exists in the soil profile but where a high water table exists at, or near, the ground surface. Some limestone areas are especially hazardous for use as dam sites. They may form tunnels, cracks or channels in the limestone below the ground surface, which are not visible from the surface. These lines of seepage may drain a dam in a very short period of time. In addition, soils in these areas are often granular. The granules do not break down readily in water and the soils remain highly permeable.

2.1 SOIL TESTING

Without extensive investigations and laboratory tests, it is difficult to recognise all of the factors that might make a site undesirable. One of the best guides to a site's suitability is the degree of success experienced by neighbours with similar dams. The absence of a layer of relatively impervious material over a storage area does not necessarily mean that the site must be abandoned. It usually means that the area will have to be treated by one of the several methods described in Section 6 under the heading 'A dam leak'. However, any of these methods may prove to be expensive.

2.1.1 Foundation

The foundation includes the valley floor and its side slopes or abutments. The requirements of a foundation for an earth-fill dam are that it provides support for the overlaying embankment under all conditions of saturation and loading, and that it provides sufficient resistance to seepage to minimise loss of water. Poor foundations can lead to failure of a dam due to cracking, tunnelling, sliding, settlement, or uplift.

The soil conditions under an embankment should be investigated to ensure that the site is suitable and that a safe structure can be designed. The complexity of the foundation survey will depend upon the site conditions and on the height of the dam wall. Test holes should be taken at intervals along the centre line of the dam (Figures 2.1 and 2.2), using either a hand auger or backhoe to cut a trench to determine the nature of the soil profile. The depth and spacing of the holes should be sufficient to determine the suitability of the foundation. The location of these holes will depend on the occurrence of significant changes in the soil profile. The holes should be deep enough to identify the underlying materials that may affect the design or safety of the structure. A record, or log, of each hole or test pit should be made. This should show the location, depth and classes of materials encountered. Each test hole location should be marked on the ground for future reference, in case more detailed surveys are required.

2.1.2   Borrow pit for embankment material

Suitable excavated material needs to be found in sufficient quantities for construction of the embankment. The further the excavation is from the embankment site, the greater the placement costs. This is a good reason for having the borrow pit inside the storage area. Test holes should be made in the chosen borrow pit areas to establish whether sufficient volumes of suitable material are available.

Selected materials for construction of a dam wall must have both the strength for the embankment to remain stable, and a sufficiently low permeability (when compacted) to prevent excessive seepage of water through the dam wall.

2.1.3   Spillway site

During a storm event, it becomes necessary to bypass run-off around the embankment of a small dam through an earth spillway. For safety reasons, the spillway should be located on a natural surface and not on the fill material. Thus, auger holes should be placed along the centre line of the proposed spillway to determine the type of material that will be encountered, its erodability, and its suitability.

2.2   SITE SELECTION CRITERIA

A problem sometimes associated with selected sites is that there may be substantial depths of pervious materials (for example, 3 m). This could require a deep cut excavation for the core trench placement beneath the dam. The presence of pervious materials could also result in more expensive de-watering problems during excavations.

2.2.1   Seepage losses

Seepage losses are affected more by the prevailing groundwater conditions than by permeable soils. For example, if an excavation is cut into sandy soil, which lies well above the water table, water will tend to seep out of the excavation. On the other hand if the water table is higher than the floor of the excavation, then obviously water will seep in. However, because most storages are built well above the water table, sands and gravels are likely to be sources of seepage.

Table 2.1 Guide to site selection based on seepage loss.

Seepage loss rate (added water – L/hr)	Recommendation
<3 L/hr (50 ml/min)	Site should be satisfactory.
3–30 L/hr (50–500 ml/min)	Site should be regarded as doubtful. This indicates the need for further tests.
>30 L/hr (500 ml/min)	The site is too permeable for a dam.

Source: Information from Nelson, 1985.
Note: L/hr = litres/hour and ml/min = millilitres/minute.

If the site is doubtful, there is a simple test often used by geo-technical people to evaluate a site. During the site exploration, several soil-sampling test holes (Figure 2.1) will be dug in the borrow pit areas under the embankment. It is important to consider the proximity of auger holes, as they can have an effect on results. These holes are usually done by hand auger (100 mm diameter and at least 3 m deep). Several holes should be selected for the seepage tests. The following procedure should be used in initially testing sub-surface soils for suitability:

•   After excavating auger holes, carefully fill each hole to two-thirds its depth with water so as to saturate the sides of the sub-surface soils. It is important not to disturb the wall of the hole during filling. If it is necessary to support the sides of the test hole, this can be done by filling gravel around a slotted pipe.
•   When water fills the auger hole to two-thirds its total depth, the depth to water level below the ground surface should be maintained in each hole. The quantity of water required, and the time intervals when water is added to maintain this depth, will need to be recorded.
•   It is important that the trials be carried out over a number of hours and that the effect of seasonal variation on moisture content are considered in the end results.

A tried and proven method of assessing the results of field testing for seepage is given below. This procedure is also used in the selection of effluent line sites for sewerage systems.

The recommendations listed in Table 2.1 are suggested guides for site selection, based on seepage loss where the auger hole has a diameter of 100 mm, a depth of 3 m and is filled 2 m deep of water (that is, the water level is 1 m below the ground surface).

If seepage loss rates greater than 3 L/hr (50 ml/min) are confirmed, then other factors must be considered before proceeding with the project. These would include the purpose of the water supply scheme, the availability of water for the dam and the likely cost of treating the seepage area.

A comparative evaluation of falling water levels over an area can then provide an indication of permeability and may indicate relative clay contents. Infiltration rings are a more sophisticated way of assessing infiltration capacity for irrigation design purposes.

Like sands and gravels, some jointed formations of rock are permeable and encourage seepage. Dams have failed due to sinkholes developing in the floor of the storage.

This usually occurs when limestone underlies the soil. Water from the storage dissolves the rock to form vertical holes, which in turn lead to underground cavities and springs.

2.2.2   Stability of dam sides

As the level of stored water rises, so does the water table in the sides of the dam. Soils and rocks, which are quite stable when dry, may become weak when saturated. This could cause a landslip, which in turn will reduce the capacity of the storage. Frequently this problem is aggravated by cutting the borrow pits too close to the reservoir sides.

2.2.3   Sedimentation in dams

Sedimentation is a problem that occurs in catchments with active soil erosion. In these cases, advice should be sought from a soil scientist. Sometimes it is possible to remedy the problem at its source; if not, it may be necessary to resort to filter strips and silt traps. Filter strips are dense stands of stiff, long-stemmed plants intermingled with grass. These strips reduce the velocity of the water and so cause silt to be deposited.

In the case of choosing a gully site where sedimentation may be a problem, the following topographical features should also be borne in mind.

- The storage should be located in a wide valley just upstream of a narrow gorge. This will initially provide maximum storage for minimum earthworks. Over time, silting will reduce the storage capacity of the dam and it will require maintenance (see Section 6.3.8).
- It should be located on the flat slope of a stream rather than on the steep slope. This provides a larger capacity of stored water for any given wall height of a dam.

2.3   FOUNDATION MATERIALS

It is possible to construct an earth-fill embankment on a suitable foundation if this has been thoroughly investigated and the design and construction procedures are adapted to site conditions. Some foundation conditions require construction measures that are relatively expensive which, in the case of small small dams, cannot be justified. Sites with such foundation conditions ordinarily should be abandoned.

The best foundation comprises, or is underlaid by, a thick layer of relatively impervious, consolidated material, which occurs at a shallow depth. Such foundations cause no stability problems. Where a suitable layer occurs at the surface, no special measures are required. It is sufficient to remove the topsoil, and scarify or disc plough the area to provide a bond with the material in the dam wall. A compacted clay cut-off trench can be constructed to extend from the surface of the ground into the impervious layer. This prevents possible failure by tunnelling and excessive seepage.

A detailed investigation should be made where the foundation consists of either pervious sand or a sand–gravel mixture, and the impervious clay layer is beyond the reach of equipment. While such a foundation might be satisfactory in terms of stability, corrective measures will be required to prevent excessive seepage and possible failure.

In the case of a foundation consisting of, or underlain by, a highly plastic clay or unconsolidated material, very careful investigation and design is required in order to obtain stability.

Water stored on bedrock foundations rarely gives cause for concern unless the rock contains seams, fissures or crevices through which water may escape at an excessive rate. Where rock is encountered in the foundation, careful investigation of the type and physical properties of the rock is required.

Foundations must be capable of supporting the weight of the dam and must be sufficiently watertight to prevent seepage under the dam. Springs, soaks or landslips indicate unstable soil conditions and should be avoided.

Therefore, the three main kinds of foundation material are:

Clay	Clay foundations are usually satisfactory, provided they are of the same material as that placed in the earth bank. However, if they are soft and saturated it may become necessary to remove them or place additional stabilising fills. Highly expansive clays, which shrink and swell during cycles of wetting and drying, may be unsuitable because of risks associated with tunnelling and high seepage rates.
Rock	Most rock can support the weight of the dam. Care must be taken to ensure that seepage does not occur between the rock foundation and the earth-fill dam so that weathering of the rock does not lead to weakening of the foundation, or that permeable zones are not created by joints and faults. Care should also be taken where expansive rock is being excavated, since elastic recovery (or expansion) of rock material occurs in response to reduced pressures, as the overburden is removed.
Sands and Gravels	The problem with this type of foundation is high seepage losses. While it is possible to build dams with these materials, the cost is frequently prohibitive. Such sites are best avoided and an alternative location found.

2.4 EMBANKMENT MATERIALS

Soils placed in dams must fulfil two conditions; they must be sufficiently impervious to keep the seepage at a safe rate, and they must have sufficient strength to ensure stable side slopes.

There are three kinds of gully dams: homogeneous, zoned, and diaphragm (see Section 3.1.2, Figure 3.2). The homogeneous dam is built from one type of soil and is the most common kind in Australia. A zoned bank consists of a centre clay core with pervious material on either side. It is considered the most stable form of small dam. The diaphragm dam is built when there is only a limited amount of clay available at the site. The bulk of the bank is constructed from relatively pervious material with a thin layer (that is, a diaphragm) of clay on the upstream slope. This layer varies from 0.6 to 1.0 m thick, depending on the height of the dam.

Good, impervious material contains about 25 per cent clay with the balance made up of silt, sand and some gravel. Too much clay results in the embankment being weak and prone to expansion and contraction with changes of moisture content. Insufficient clay can cause excessive seepage through the bank.

The usual method for exploring the material at a potential dam site is hand auger boring. This is the cheapest method, although it is very hard work for the operator and provides a disturbed sample. It is therefore advisable to sink a test pit or trench so that the soil can be examined in its natural state.

Dam sites are tested on a fixed pattern. Small dams (up to 3 m high) have a minimum of six test holes, four in the centre-line (including one on the spillway) and at least two in the borrow pit area (Figure 2.2). For larger dams the number of test holes is increased, with holes at 20 m intervals in borrow pits where sites are steep or complicated. This spacing can be increased to a 70×100 m grid when the site is flat or uncomplicated. The test holes on the centre-line of larger dams are spaced at about 30 m intervals. The test holes in the borrow pits are sunk to about 3 m or to rock, while those in the dam centre-line are put down to at least three-quarters of the dam height or to rock.

When the exploration has been completed, all test holes and pits should be carefully filled to prevent human and stock injury and compacted to prevent seepage/leaks.

2.5   SITE INVESTIGATION OF MATERIALS

All the earth fill for the embankment should be excavated from within the water storage area when possible, and if required, from any cut spillway areas (SCA, 1979). At the investigatory stage possible borrow areas should be identified – initially by eye – to ascertain soil type from vegetation, visible soil, position on slope and so on. Most favourable areas can then be investigated.

Initial investigations involve the use of a backhoe to dig borrow pits and an auger to assess the condition of topsoil, subsoil and foundation in the embankment area. Auger holes dug on a grid to depths of 3 m throughout a potential source area will allow a general assessment of soil types to be made. A series of trial pits using a backhoe, can then be dug in more promising areas to allow a visual assessment of the soil profile according to soil classification techniques. Samples can be taken for texture and laboratory analysis.

2.5.1   Soil texture tests

Soil texture tests are carried out to determine soil types. The relative proportions of sand, silt and clay are used to determine the textural class of a soil. The internationally accepted and most used tool for initially identifying soils for dam building is the United States Department of Agriculture (USDA) texture diagram shown in Figure 2.3. Also included, in Table 2.2, are the texture classes as defined by a percentage for sand and clay.

2.5.2   Unified soil classification

Materials are classified according to the 'Unified Soil Classification' system. This system of classification uses grouped symbols to indicate the physical properties of the material. The system is based on the size of the particles, the relative amounts and the characteristics of the very fine particles.

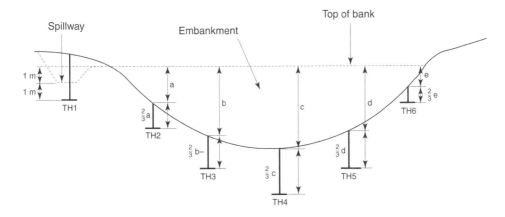

Figure 2.1 Depth of test holes along centre-line (*Source*: WAWA, 1991).

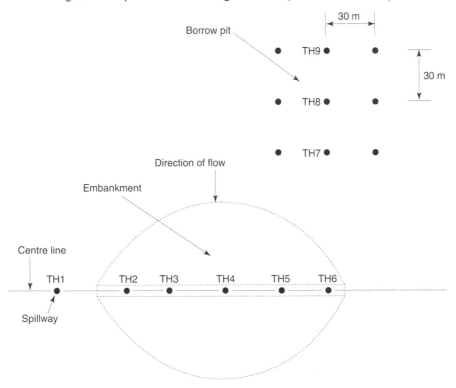

Figure 2.2 Test hole plan for a small dam (*Source*: WAWA, 1991).

Table 2.2 provides the list of grouped symbols used under the 'Unified Soil Classification', with their common names, organised in order from largest to finest particles eg. gravels, sands, silts, clays and organic material.

To define the textural classification of soils precisely requires laboratory techniques. However, with experience and specific local knowledge, hand testing to

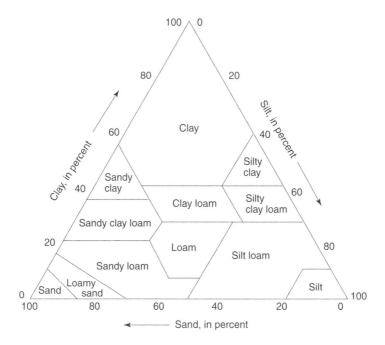

Figure 2.3 Textural classification chart (*Source*: SCS USDA, 1967).

Table 2.2 Textural Classes.

Textural Class	Clay %	Sand %	Silt %
Clay	>40%	<45%	<40%
Sandy Clay	35–55%	45–65%	<20%
Silty Clay	40–60%	<20%	40–60%
Clay Loam	25–40%	20–45%	15–55%
Silty Clay Loam	25–40%	<20%	40–75%
Sandy Clay Loam	20–35%	45–80%	<30%
Loam	10–25%	25–55%	30–50%
Silty Loam	<25%	<50%	50–90%
Sandy Loam	<20%	45–85%	<50%
Loamy Sand	<15%	70–90%	<30%
Silt	<10%	<20%	>80%
Sand	<10%	>85%	<20%

Source: Lewis

determine texture can prove important in the initial stages of identifying appropriate earth fill materials. Clay soil sites (that is, soils with a higher percentage of clay) can be defined in the field. These soils are used for the core and upstream batter of the embankment. Silts are often similar in both appearance and feel to wet clays when dry, but can usually be differentiated when wet as the clay will exhibit sticky, plastic-like characteristics while silt has a silky, smooth feeling with a tendency to disperse.

Table 2.3 Unified Soil Classification.

Texture class	Soil Group	Soil characteristics
GRAVEL	GW	Well-graded gravels; gravel-sand, little or no fines
	GP	Poorly graded gravels; gravel-sand mixture, little or no fines
	GM	Silty gravels; poorly graded gravel-sand-silt mixture
	GC	Clayey gravels; poorly graded gravel-sand-clay mixture
SANDS	SW	Well-graded sands; gravelly sands, little or no fines
	SP	Poorly graded sands; gravelly sands, little or no fines
	SM	Silty sands; poorly graded sand-silt mixture
	SC	Clayey sands; poorly graded sand-clay mixture
SILTS	ML	Inorganic silts and fine sands; rock flour, silt or clayey fine sands with slight plasticity
	CL	Inorganic clays low medium plasticity; gravelly and sandy clays; silty clays; lean clays
	MH	Inorganic silt; micaceous or diatomaceous fine sandy or silty soils; elastic silts
	CH	Inorganic clays of high plasticity; fat clay
	OH	Organic clays of medium to high plasticity
	Pt	Peat and other highly organic soils

Source: SCS USDA, 1967.

Hand-testing techniques involve taking a small sample of a soil – usually in the hand not required for making notes – dampening it (avoiding soaking it), and rolling it into a ball to examine its cohesive constituents. A good clay can be manipulated into a thin strip without breaking up, rolled into a ball and dropped onto a flat surface from waist height without cracking unduly, and cut to exhibit a shiny, smooth surface.

2.6 ANALYSIS OF SOIL

Analysis of soil samples should be carried out to assess components and compaction characteristics, and to check for other factors that may make apparently good soil unsuitable. Correlation of these results with previous work will allow estimates to be made of available earth fill, overburden to be removed and unsuitable areas to be avoided.

The importance of a correct analytical approach to determine the various soil types for a zoned embankment cannot be over-emphasised. Although using a soil laboratory is expensive, the results can more than repay the cost involved and, more often than not, will ensure the exclusion of doubtful material in the construction process (Stephens, 1991).

2.6.1 Core trench

The most satisfactory way of dealing with permeable layers beneath an embankment is to install a core trench (see Section 3.1.2). The core trench must be excavated through any permeable layers and keyed into impermeable base material, then backfilled with compacted clay. Compaction is important, particularly if the in-situ clay is well structured, when compaction must be good enough to break down the permeable structure as well as to remove the air introduced to the material during excavation. If not compacted, initial loss rates will be high but later consolidation under the pressure of the water in storage may reduce the loss. Where no clay is available, or more usually where the available material contains insufficient or unsatisfactory clay and is therefore permeable, it may be advantageous to import suitable alternative material (see Section 6.4.2)

2.6.2 Embankment soil

For embankment, the properties of soils in the undisturbed state require special examination. Most important are the denseness of coarse grained soils and the consistency of the fine grained soils. Also the consistency of the same soil remoulded with its natural water content should be observed. For example, clay may be described as "stiff and brittle in undisturbed state and very soft and sticky in remoulded state".

A material is permeable if it is capable of allowing the passage through it of another substance, usually a liquid. For a material to be permeable it must firstly be pervious, or contain voids and secondly at least some of the voids must be continuous or interconnected. That is why placement of materials within embankment is so important.

2.7 LOCATION OF SOIL

Small dams are nearly always built with the material immediately available on site. A good preliminary guide for selecting a source of material that is suitable for building a stable earth dam is the behaviour of other dams built in the neighbourhood. In an initial investigation it is sufficient to test that the material is suitable in quantity and quality by putting down test holes or test pits. The availability of materials is most important. Having determined the particular group to which a soil belongs, it is necessary that we recognise the factors which control the properties of the soils within the group. Engineering properties of prime importance in determining the design of small earth dams and selection of soils are permeability, workability, shear strength, compressibility and density.

Clays   Clay soils are very fine grained soils of colloid size. A soil with more than approximately 50% clay content is defined as a clay soil (see Figure 2.3). They are essentially plastic cohesive soils which shrink on drying, expand on wetting, and when compressed give up water (dilatancy). They can be defined as having low liquid limit (less than 50%) or high liquid limit (greater than 50%). A clay of low liquid limit will have slight to no dilatancy, medium dry strength and medium toughness; whereas a clay of high liquid limit will have no dilatancy, high strength and will be very tough.

When constructing a dam consideration needs to be given first to the placement of soils with high clay content within a core trench to minimize seepage flow line through the embankment (see Figure 3.2).

Most small dams are built on clay foundations but care must be taken to ensure that the clay is of a quality such that, when saturated, it will not settle excessively under the superimposed load. For embankments up to about 10 metres high, most of the inorganic materials will have sufficient strength. Occasionally a dam is built on weak clay foundation when the contractor has not been aware of the hazard; the result is an upward bulge of the ground surface downstream from the centre of the dam, subsidence of the dam in the centre, and perhaps subsequent failure.

Once clay content has been established from field test boreholes then further testing is needed to establish if it has good water holding qualities

Silts   Silts consist of non-plastic fines and should be avoided in any embankment earthworks. This is because they are inherently unstable in the presence of water and have a tendency to become "quick" when saturated. Silt masses undergo change of volume with change of shape (the property of dilatancy) in contrast to clays which retain their volume with change of shape (the property of plasticity). Silt can have either low liquid limit or high liquid limits. Silts with low liquid limit will have medium to quick dilatancy, almost no dry strength and no toughness. Silts with high liquid limit will have slight to no dilatancy, slight to medium dry strength and slight to medium toughness.

Sands   In general, a soil with high sand content should not be used in embankment earthworks, however, it can be used in downstream batter and in the treatment of seepage flow lines for internal drainage (see Figure 3.2). It should not be used elsewhere unless there is no alternative. Although it has apparent cohesion when damp, it has no cohesion when dry or saturated.

2.8 UNSUITABLE MATERIAL

There are many materials that are not suitable for dam construction. While clay is one of the most useful soils for dam building, silt, for example, when wet, can be troublesome. Silt and clays may appear similar and may exist in the same size ranges on any grading chart, but behave very differently in the presence of moisture. Generally silts are very hard to compact. Changes in the moisture content after compaction can substantially alter the properties of the material in an embankment causing shrinkage and subsequent swelling on re-wetting. It can change the in-situ shear strength and compressibility characteristics and any or all these changes can be a potential source of trouble unless closely watched.

Other materials to avoid include: vegetable matter in topsoil stripped from the dam site or borrow pit area; expansive clays; dispersive soils such as calcitic and sodic soils, calcitic soils because of their high porosity, sodic soils because of their lack of cohesion when wet.

Section 3

Design

Australians spend millions of dollars each year on the construction of small dams for rural communities. This money is wasted because a large proportion of these dams fail. The solution to this problem rests in more thorough investigation and improved standards of design and construction. Unfortunately such proposals are often regarded with apprehension because many people consider that higher standards mean increased costs (Lewis, 1995a).

The combined cost of field investigation and design usually represents only 5 to 10 per cent of the total capital cost of the dam. Furthermore, if the dam is constructed to a well-designed plan many cost-saving features can be incorporated. So you may well ask 'Can I afford the extravagance of a cheap dam?'

3.1 ITEMS THAT NEED TO BE CONSIDERED

An earth-filled dam of any size must be designed so that it is structurally sound and stable during its operational life. They are simple structures that rely mainly on their mass to resist sliding and overturning. Modern construction techniques and developments in soil mechanics have greatly increased the safety and durability of these structures.

The following design considerations must be met if safety is to be maintained:

- the batter slopes must be stable and resistant to movement under different operating conditions, including rapid drawdown.
- the batter slope on the upstream side must be able to handle across-water wave action.
- the earth embankment must be safe from overtopping by both flood inflow and wave action. This is the reason for providing a 1 m freeboard from full supply level to the top of the crest.
- seepage through the embankment and beneath the foundation needs to be controlled to prevent piping or tunnelling along a line of weakness through, or under the dam.
- the embankment needs to be protected from a major failure due to earthquake, if the site is known to be earthquake-prone.

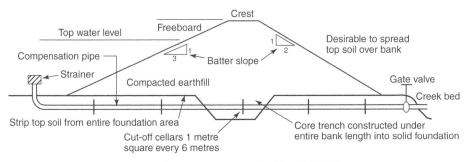

TYPICAL EMBANKMENT CROSS-SECTION

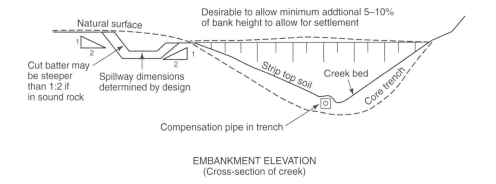

EMBANKMENT ELEVATION
(Cross-section of creek)

Figure 3.1 Cross-section and elevation view (*Source*: SRW, 1995).

3.1.1   Embankment types

i Homogeneous

Historically, farm dam embankments across drainage lines and waterways were constructed of materials of uniform consistency, for example, clay. The embankment should contain a minimum of 30% clay with the balance made up of other materials such as silt, sand and gravel.

Soils used in earth dams should also be sufficiently impervious to reduce seepage to a safe rate and insure stability within batter slopes, upstream and downstream.

When dams are to be built from one type of impervious material, then a homogenous bank can be constructed. Normally homogenous dams are confined to relatively small heights. It is recommended that dams in excess of 6 metres that a zoned embankment be constructed (see Figure 3.2) particularly when there is a likelihood that there will be a rapid drawdown in the storage.

The upstream and downstream slopes adopted are generally in accordance with slopes given for various groups of soils in Table 3.1.

This method of construction is simple, straightforward and suitable for those sites where there is sufficient suitable material. Protection from seepage and slipping is provided by flattening the downstream batter, that is from a 2:1 to 3:1 gradient (horizontal to vertical), and providing a thick covering of topsoil to carry any seepage to the toe of the bank.

Figure 3.2 illustrates this situation and shows the incorporation of a processed drain material (sand/gravel or similar).

Embankment materials vary widely from place to place, particularly in respect to gradation and permeability. If the difference in permeability between the impervious core and the downstream batter is great, no internal drainage is required. If the variation of the permeability between the inner core and the batter is not sufficient, then the embankment will become saturated after prolonged storage at full supply. Consequently, the downstream slope will show seepage to a height of approximately one-third the depth of the high water level. Such saturation reduces the stability of the dam and creates maintenance problems. This pressure line is also shown in Figure 3.2 and is called the 'phreatic' water surface line. Major maintenance problems can occur if this is not taken into account. These problems include undesirable growth of vegetation on the batter and minor slippage from trafficking by animals and vehicles.

ii Zoned

At those sites where there are varying soil materials with widely differing construction properties, and where high dams are being considered, a zoned embankment should be considered. Zoned embankments consist of an impervious core/blanket held in place and protected by a more pervious 'shell'. The core may be centrally placed, sloping or placed on the upstream slope in the form of a blanket (Figure 3.2).

A zoned embankment consists of a central clay core with pervious material forming a shell on each side.

The most stable earth dam that can be built is the zoned dam; it should always be the preferred option provided that the right kinds of materials are available. It has the advantage of

- placement of a choice of pervious or impervious materials on batter slopes
- permitting the construction of larger dams
- allowing steeper batter slopes therefore reducing the volume of materials to be excavated
- a better opportunity for using all suitable material on site and within the storage.

Site selection and transporting materials can make this treatment uneconomical and alternatives may need to be used (see Section 6.4.2).

The use of pervious (sandy) soil materials in the shell can greatly reduce the volume of the embankment as steeper batters can be used. Upstream blankets are commonly used in conjunction with lining of the storage area of the dam.

For high embankments (greater than 10 m) the use of toe drains and blanket drains allows the downstream slope to be maintained at a reasonable slope of approximately 3:1 (horizontal : vertical). Without these provisions, much flatter slopes would be required to maintain slope stability, particularly for clayey soils.

iii Artificial liners

Synthetic materials such as plastic liners are used to seal farm dams where other methods are unsuccessful or too expensive. Situations where this is likely to be the case are friable clay soil or coarse sand, but these liners can be susceptible to being punctured by vermin if unprotected with a soil blanket (see Figure 6.11). In Section 6, details are given on other alternatives that can be used to minimise leaks in dams. These materials are currently being tried in new dams in areas prone to problems.

3.1.2   Core trench

Under the middle of the dam a core trench must be provided to minimise seepage beneath the dam wall. The core trench should have side slopes no steeper than 1:1 (horizontal to vertical) for a depth up to 3 m and no steeper than 1.5:1 (horizontal to vertical) for greater depths. The bottom width of the trench should be 1.5 times the height of the dam, or a minimum width equal to the width of a bulldozer or scraper. The depth of the core would generally extend to bedrock, or to an impervious stratum of soil sufficient to prevent excessive seepage beneath the dam.

The core trench material should be placed in layers with a maximum thickness of 100 mm. It is important that every layer is well compacted. The whole dam length should be completed at the one time. If this is not possible, then each section must be well keyed and bonded to the next. To avoid seepage and structural problems, the core trench and embankment shell should be designed as one homogenous unit. Any water collected in the core trench should be removed before backfill operations are started.

Occasionally a dam is built on a weak clay foundation or permeable layer resulting in subsequent failure. The most satisfactory way of dealing with this problem is to initially install a clay core to enhance structural integrity. In addition, to reduce seepage losses beneath an embankment and enhance stability it is usual to use a cut off trench which intercepts all permeable material down to a thick impermeable material or to parent bedrock.

Where the foundation consists of pervious materials at or near the surface, with rock or impervious materials at a greater depth, seepage through the pervious layer should be reduced to prevent piping and excessive losses. Usually a cut-off joining the impervious stratum in the foundation with the base of the dam is needed.

The trench should be filled with successive, thin layers of relatively impervious material, with each layer being thoroughly compacted at near optimum moisture conditions, before the next layer is placed.

3.1.3   Embankment batter slope

For homogenous dams, the following batter slopes are recommended for embankments built of soils classified according to the Unified Soil Classification system (see Section 2.5.2).

The batter slopes for most embankments on strong foundations can be 3:1 (horizontal to vertical) upstream and 2:1 (horizontal to vertical) downstream. However, flatter batter slopes should be considered for dams structured on inorganic clay or highly plastic and very fine inorganic silt. Organic soils are not useable as an embankment material. They tend to be placed on the outside batters to establish grassy vegetation. They act as a blanket by reducing internal dam moisture losses.

The batter slopes of a dam depend primarily on the stability of the material in the embankment. The greater the stability of the fill material, the steeper the slopes

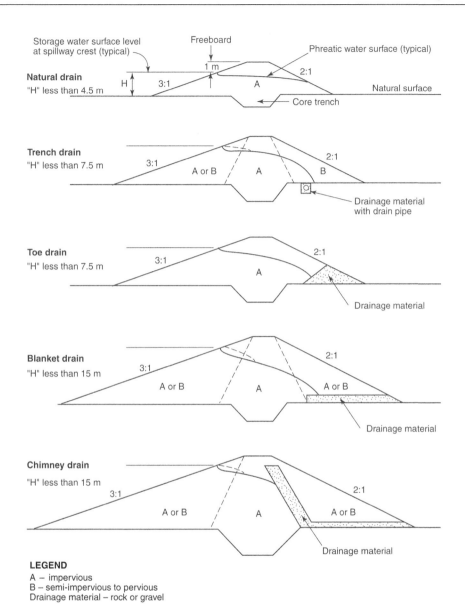

Figure 3.2 Zoned dams with treatment of seepage flow lines for internal drainage
(*Source:* modified from CDWR, 1986).
Legend:
I – impervious zone, which under the Unified Soil Classification would be used for soil types, GC, SC, CL and CH.
II – semi-impervious layer, which under the Unified Soil Classification would be used for soil types, GW, GP, gravelly SW, and gravelly SP.
Note: Batter gradelines on upstream and downstream sides would need to be 3:1 (horizontal to vertical) and 2:1 (horizontal to vertical) respectively. This is to accommodate for rapid drawdown of the water level. Gradelines are totally dependent on soil types and need to be carefully tested and checked before adoption. Each of the cross-sections (top to bottom) in Figure 3.2 is based on a bank height of less than 4.5 m, 7.5 m, 7.5 m, 15 m, and 15 m respectively.

Table 3.1 Batter slope with soil classification and height.

Case type	A Homogeneous or modified				B Modified homogeneous			
Purpose	Detention or storage				Storage			
Subject to rapid drawdown	No				Yes			
Soil classification Dam Height (m) & Slope	GW GP SW SP	GC GM SC SM	CL ML	CH MH	GW GP SW SP	GC GM SC SM	CL ML	CH MH
0–3 U/S	P	2.5 : 1	2.5 : 1	3.5 : 1	P	3 : 1	3.5 : 1	4 : 1
D/S		2 : 1	2 : 1	2.5 : 1		2 : 1	2.5 : 1	2.5 : 1
3–7 U/S	P	2.5 : 1	3 : 1	3 : 1	P	3.5 : 1	4 : 1	4 : 1
D/S		2.5 : 1	2.5 : 1	3 : 1		2.5 : 1	3 : 1	3 : 1
7–10 U/S	P	3 : 1	3 : 1	3.5 : 1	P	3.5 : 1	4 : 1	4 : 1
D/S		3 : 1	3 : 1	3 : 1		3 : 1	3.5 : 1	3.5 : 1

Note:
U/S = upstream slope & D/S = downstream slope.
rapid drawdown rates of 1000 mm or more per day may follow prolonged storage at high water level.
P (pervious) denotes soils which are not suitable. OL and OH soils are not recommended for major portions of homogeneous earth-fill dams. Pt (Peat) soils are unsuitable.
See Table 2.3 for soil groups.

may be. The more unstable materials require flatter slopes. Table 3.1 contains recommended maximum slopes for the upstream and downstream faces of dams constructed of various materials.

3.1.4   Crest width

The crest width should be increased as the height of the dam increases. The generally accepted empirical formula for crest width is:-

$$\text{Crest width (m)} = H^{0.5} + 1$$

where H is the height of the crest of the embankment above the bed of the gully in metres.

The minimum crest width should be about 2.5 m irrespective of H, so that machinery can work on the crest.

Table 3.2 contains recommended top widths for embankments of various heights based on above formula.

Where the top of the embankment is to be used as a road access track then the top width should provide for a shoulder on each side of the roadway. The crest width in such cases should not be less than 4 m. An exception to this is a turkey's nest dam where usually the embankment is not high, and filling of the storage (by pumping) is under control. Since there are no hazards, the crest width may be reduced to as little as 1.5 m, if necessary (and if plant is available to work or trim such a narrow crest).

Table 3.2 Height to crest width.

Dam height (m)	Crest width (m)
3	2.75
4	3.00
5	3.25
6	3.50
7	3.65
8	3.85
9	4.00

Source: SCA, 1983.

3.1.5 Freeboard

Care must be taken to allow for wave action, especially on dams that have 0.7 km or more of exposed water surface. For less than 0.7 km see Figure 3.4. Erosion by wave action has certainly occurred on some of the large 'shallow storages' and on large ring tanks. It is accentuated on very friable soils and/or fast slaking clays. It is important to establish good grass protection on the upstream batter (as difficult as this might be with changing water levels) to minimise wave erosion. Rock-beaching would be the conventional engineering solution to wave erosion. However, this is too expensive for most farm dams. As a cheaper and more risky alternative – trusting the landowner to do everything possible to establish the necessary grass cover, and to replace eroded material when necessary – is recommended.

Freeboard is the added height of the dam provided, as a safety factor, to prevent waves or run-off from storms greater than the design allows from overtopping the embankment. It comprises the vertical distance between the elevation of the water surface in the dam, when it is full, and the elevation of the top of the dam after all settlement has taken place. A large number of dams have failed due to overtopping and consequently greater attention must be paid to this feature. Freeboard should not be less than 1.0 m and should include provision for:

a   depth of flood surcharge to pass water through the spillway;
b   b wave action, which can be calculated from Hawksley's formula:

$$H = 0.0138\sqrt{F}$$

where H = wave height in metres,

F = fetch distance (the longest exposed water surface in metres).

See Figures 3.3 and 3.4;
c   n additional allowance of 0.3 m for unevenness in the crest level.
     Using these provisions:
     Freeboard (m) = depth of flood surcharge + wave action + 0.3.

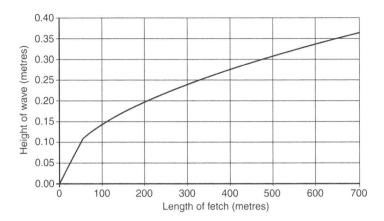

Figure 3.3 Wave action.

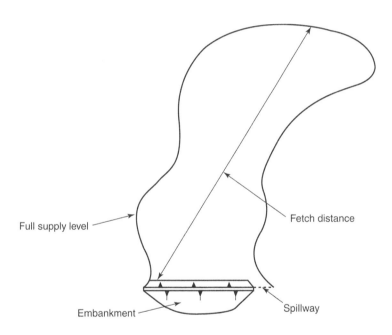

Figure 3.4 Wave action based on fetch distance across the storage.

3.1.6 Alternative ways of batter protection

A farm dam should not be considered complete until proper protection from erosive wave action, livestock and other sources of damage has been provided. Dams that lack such protection may be short-lived, and the cost of maintenance is usually high. In most areas, the exposed surfaces of the dam, spillway, borrow areas and other disturbed surfaces can be protected against erosion by establishing a good cover of

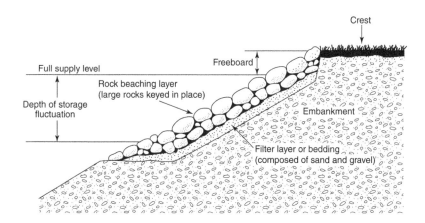

Figure 3.5 Typical rock-beaching (*Source*: modified SCS USDA, 1969).

sod-forming grass. Occasionally, there is a need for better protection against wave action than will be provided by grass cover. Methods used to provide this protection include earth berms, log booms, and rock riprap.

i Berms

A berm, 2.5 to 3 m in width and located at normal dam level, will often provide adequate protection from wave action. The face of the dam above the berm should be protected by vegetation.

ii Floating booms

A boom may consist of a single or double line of pine logs chained together and securely anchored to each end of the dam. The boom should be tied end-to-end as close together as practical. There should be enough slack in the line to allow the boom to adjust itself to fluctuating levels in the dam. Double rows of logs should be framed together to act as a unit. The boom should be placed in order to float about 1.5 m upstream from the face of the dam for best results. In the case of a curved dam, anchor posts may be required at the ends in order to prevent the boom from riding on the slope. Booms afford a high degree of protection and are relatively inexpensive, especially in areas where timber is readily available.

iii Rock-beaching

Where the water level in the dam can be expected to fluctuate widely, or where a high degree of protection is required, the use of rock-beaching is a most effective method of control (Figure 3.5). Rock-beaching should extend from the top of the dam, down the upstream face to a level at least 1 metre below the lowest expected level of the water in the dam. Rock-beaching may be placed by machine or by hand. Machine placing requires more stone but less labour. The layer of stone should be durable and large enough not to be displaced by waves. Where rock-beaching is not continuous with the upstream toe, a berm should be provided on the upstream face to support the layer of

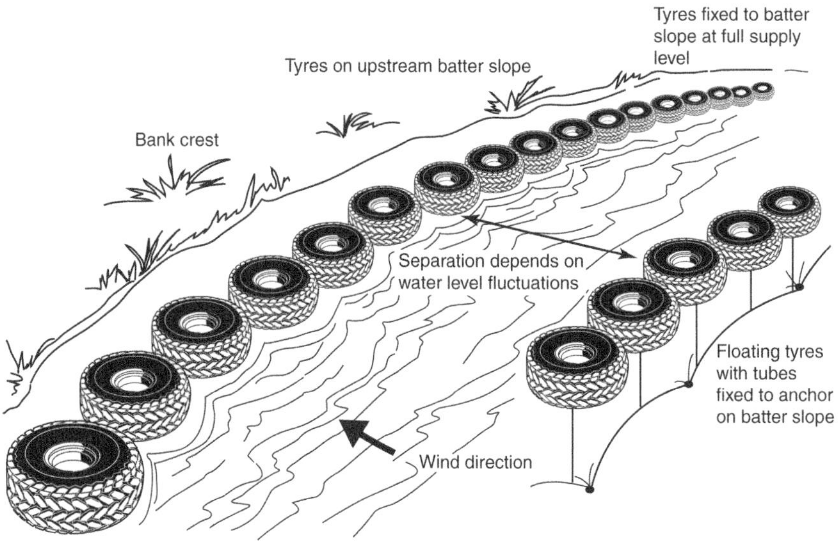

Figure 3.6   Floating tyre systems (*Source*: Floating Tyre System, Pty Ltd).

rock-beaching. In some circumstances graded gravel filters may be required under the rock-beaching layer.

iv   Other methods

Other methods include increasing the crest width of the dam, flattening the upstream slope of the embankment, and applying a layer of coarse sand and gravel on a 10:1 (horizontal : vertical) slope. These methods are applicable to arid areas where vegetation is not reliable and rock and timber is not readily available.

Old tyres tied together and secured to the dam face have also been used to maintain vegetation cover (Figure 3.6; see also Section 3.1.8 on fencing).

3.1.7   Topsoil cover

Prior to the placement of topsoil onto the compacted embankment, the surface should be roughened to assist in combining the different soil types. Topsoil must be placed over the entire embankment to a depth of at least 100 to 150 mm and grassed with a good holding grass. The purpose of the topsoil cover is to:

- reduce surface erosion on either side of the batter slope;
- minimise surface cracking in the embankment;
- lessen the tendency of the surface material in contact with storage water being dispersed; and
- lessen wide fluctuations in embankment moisture content.

3.1.8 Fencing

The complete fencing of embankment type dams is recommended where livestock are grazed or fed in adjacent areas. Fencing provides the protection needed to develop and maintain vegetation cover. When combined with a water facility below the dam, fencing allows good quality drinking water and eliminates the danger of pollution by livestock. Fencing also improves wildlife habitat.

3.2 FLOOD FLOW ESTIMATION

All dams formed by embankments across natural drainage ways require the protection of a carefully designed spillway or a combination of spillways. The function of outlet structures is to pass storm run-off around or under the embankment to prevent over-topping. In addition, the structures must convey water from the dam safely to a stable outlet below, without damaging the downstream slope of the embankment. Spillways can consist of more than one component: trickle pipes, drop inlet structures, chute or earth spillways (see Section 3.3).

Spillway design dimensions are linked to the size and the character of the catchment. A catchment with rocky or steep surfaces (and therefore high run-off) will have higher peak floods than a catchment of the same area, within the same climatological zone and with flatter, well-vegetated slopes. Similarly, a long narrow catchment will have a greater time of concentration of floodwater after a rainstorm than a broad catchment with the same area, climate, gradient and vegetation.

The peak flood is the maximum flood to be expected from a catchment following a rainfall of estimated intensity and duration for a selected return period. In many parts of the world information is not available or smaller streams are not gauged to allow estimation of such floods for spillway design purposes. It is economically prudent to study the hydrology, climate, and topography of a catchment supplying run-off to a large dam in order to determine reasonably accurate estimates of peak floods. However, for smaller dams, unless this information is already available, the engineer can rarely justify the cost of such an exercise and must resort to other means to estimate the maximum possible flood.

Where no other data is available, a very approximate peak flood estimate can be made by taking the highest daily rainfall figure for the catchment and making the assumptions that all dams in the same catchment are 100 per cent full, the ground is saturated, and 100 per cent run-off will occur.

An important element in designing spillways of a dam is to establish run-off within a specified return period (recurrence interval). Selection of a return period depends on the economic balance between cost of periodic repair or replacement and the cost of providing additional capacity to reduce the cost of repair or replacement. In some instances, the possibility of causing damage to other works by a failure may dictate the choice of the return period.

To protect a dam, flood flows entering the storage must be able to pass around the embankment by a spillway or, as it is sometimes called, a by-wash. Most spillways on farm dams are cut into the earth because concrete is too expensive. They are generally designed to pass a 2 per cent probable flood. This refers to a flood so large that there is only a 2 in 100 chance of it occurring in any year.

The generally accepted flood frequency return periods used are:

Minor dams (depending on consequences of overtopping)	10–50 years
Detention dams	50–100 years
Large dams	>100 years

3.2.1   Peak flow estimation

The wide range of variation of Australian climatic conditions makes it difficult to provide a simple method for estimating flood flows that can be applied generally for small to medium-sized rural catchments. The methods more often used are based on *Australian Rainfall and Run-off, A Guide to Flood Estimation* (IEA, 1987). For an understanding of the methods discussed and selection of appropriate design values, it is important that the probabilistic nature of the methods is recognised.

Engineers can, in most cases, calculate peak discharge for flood flow conditions based on available observed flood flow data in the region of interest. For most of Australia, sufficient data is available for the derivation of methods that reasonably reflect reality. The methods in *Australian Rainfall and Run-off (ARR)* take into consideration rainfall intensity, catchment characteristics and size, average slope of the waterway and its length from source to the dam site. These same methods of estimating peak discharge are used by local government and drainage and rural water authorities for the design of most structures on waterways.

Estimation of required spillway size for farm dams is given in Section 3.3.3. This is based on design discharge for small to medium rural catchments for an Average Recurrence Interval (ARI) of 50 years, based on *ARR* (*ibid.*). This has been calculated in Table 3.3 (Lewis, 2001a). The information in the tables should only be used as a guide for assessment. An engineer experienced in hydrology and farm dam design should be consulted before any works are commenced.

The formula as used in design of Table 3.3 is:

$$Q_Y = 0.278 C_Y \cdot I_{tc \cdot Y} \cdot A$$

where Q_Y = peak flow rate (m^3/s) of average recurrence interval (ARI) of Y years

C_Y = run-off coefficient (dimensionless) for ARI of Y years

$I_{tc \cdot Y}$ = average rainfall intensity (mm/h) for design duration of t_c hours and ARI of Y years

A = area of catchment (km^2). If area is in hectares instead of km^2, the conversion factor is 0.00278 (or 1/360).

3.3   Outlet structures

There are various types of components that are incorporated into an embankment to allow for the passing of water flows, that is, storm run-off, compensation flows or environmental flows under a licence condition. Some of the common ones are explained in the rest of this section. The tables and information provided should only be used as a guide in the selection of sizes of pipes and spillway dimensions. It should

Table 3.3 Typical design discharge data for 50-year ARI.

Catchment area (ha)	Rainfall intensity (mm/hr)	"C_Y" Design run-off coefficient contours (%)						
		0.05	0.10	0.15	0.20	0.25	0.30	0.35
		"Q_Y" Spillway design rainfall intensity (m^3/s)						
50	65	0.5	1.1	1.6	2.2	2.7	3.4	3.8
100	60	1.0	2.0	3.0	4.0	5.0	6.0	7.0
150	55	1.4	2.8	4.1	5.5	6.9	8.3	9.6
200	50	1.7	3.3	5.0	6.7	8.3	10.0	11.7
250	48	2.0	4.0	6.0	8.0	10.0	12.0	14.0
300	46	2.3	4.6	6.9	9.2	11.5	13.8	16.1
350	44	2.6	5.1	7.7	10.3	12.8	15.4	18.0
400	42	2.8	5.6	8.4	11.2	14.0	16.8	19.6
450	41	3.1	6.2	9.2	12.3	15.4	18.5	21.5
500	40	3.3	6.7	10.0	13.3	16.7	20.0	23.4
750	38	4.8	9.5	14.3	19.0	23.8	28.5	33.3
1000	36	6.0	12.0	18.0	24.0	30.0	36.0	42.0
1250	34	7.1	14.2	21.3	28.4	35.4	42.5	49.6
1500	33	8.3	16.5	24.8	33.0	41.3	49.5	57.8

Source: Lewis, 2001.
Note: For details of "C_Y" Design run-off coefficient contours, see *Australian Rainfall and Run-off (ARR)–A Guide to Flood Estimation,* Volume No. 2.

be remembered that these numbers are not necessarily applicable to all cases. They are general values rather than site-specific. It is recommended that an engineer be employed to advise on these types of issues when the dam is being designed.

3.3.1 Earth spillways

An earth spillway is an earth or vegetated channel, designed to discharge the peak flow calculated for the catchment. Where catchments are small and long duration flows are not a problem, it may be feasible to handle the run-off safely with only a vegetated spillway.

Earth spillways, as discussed here, apply to both vegetated and non-vegetated spillways, the latter being used where climatic or soil conditions make it impossible to grow or maintain a suitable grass cover. Earth spillways are usually excavated, but may exist as a natural spillway on a well-vegetated saddle or drainage line. In either case, the spillway must discharge the design peak flow at a non-erosive velocity to a safe point of release. Exiting earth spillways, whether vegetated or non-vegetated, should not be built on fill material.

Earth spillway structures have certain limitations. They should be used only where the soils and topography permit safe discharge of the peak flow at a point well away from the dam at a velocity that will not cause appreciable erosion. Temporary flood storage provided in the dam has been used to reduce the design flow or frequency of use in the spillway, if a trickle pipe is incorporated in the design.

3.3.2   Design spillway capacity

Emergency earth spillways should have the capacity to discharge the peak flow from the catchment resulting from a storm. The procedure for the determining peak flood flow is presented in Section 3.2 of this book.

The definition of an excavated earth spillway consists of the four elements in Figures 3.8. These are:

- approach (entry) channel
- control section
- exit channel, and
- spill section.

Each element has a special function. The flow enters the spillway through the approach channel. The flow is controlled in the level portion and exit section and passes through critical depth at the downstream edge of the level portion. The flow is then discharged through the spill section.

The direction of slope of the spill section must be such that overflow will not flow against any part of the embankment. Wing banks may be used to direct the outflow to a safe point of release. The floor of the spillway should be excavated into original earth for the full design width. Where this is not practical, the end of the dam embankment and any earth fill constructed to confine the flow, should be protected by vegetation or riprap. It is desirable that the entrance to the approach channel be widened so it is at least 50 per cent greater than the designed bottom width at the control section (Figure 3.7). The approach channel should be reasonably short and should be planned with smooth, easy curves for alignment. It should have a slope toward the dam of not less than 2 per cent except in rock, to ensure drainage and low inlet losses.

The control section should be located near the intersection of the extended centre-line of the dam with the centre-line of the spillway. A level section, at least 6 m in length, should be provided immediately upstream from the control section.

The exit spillway must have a slope that is adequate to discharge the peak flow within the channel. A slope which is adequate for drainage is generally sufficient (for example, 0.2 per cent). The exit spillway should be straight or gently curved to spread the flow to a uniform depth over the lip of the spill section.

The spill section is the natural slope over which the peak flood-flow passes. The width of the spill section is selected so that the permissible velocities for the soil type, or the planned grass cover, are not exceeded.

3.3.3   Selecting spillway dimensions

With the required discharge capacity (Q) known (see Section 3.2) and the natural slope of the spill section determined from plotting the spillway centre-line profile, the bottom width of the control section, the width of the spill section and the depth of the flow can be selected. Tables 3.5 and 3.6 give the appropriate control section width for discharge capacity at various depths of flow. A spillway requires a level section with a minimum length of 8 m and a cross-sectional area calculated from the formula:

$$Q = 1.546BH^{1.5}$$

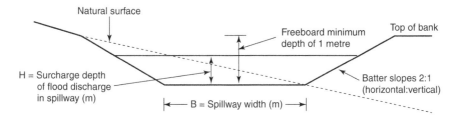

Figure 3.7 Spillway cross-section.

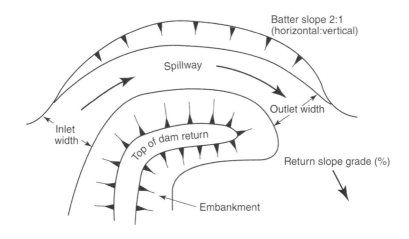

Figure 3.8 Earth spillway.

where, Q = either a 1 in 50 or 1 in 100-year peak discharge (m³/s)
B = spillway width (m)
H = surcharge depth of flood discharge in spillway (m).

Spillway side slopes should be no steeper than 2:1 (Figure 3.7) unless the spillway is excavated into rock, in which case the side slopes may be vertical. Usually, the selected bottom width of the channel should not exceed 35 times the design depth of flow. Where this ratio of bottom width to depth is exceeded, the channel is likely to be damaged by meandering flow and accumulated debris. Whenever the required bottom width of the spillway is excessive, consideration should be given to the use of a spillway at each end of the dam. These two spillways need not be of equal width so long as their total capacity meets requirements. In cases where the required discharge capacity exceeds the ranges shown (Tables 3.4 and 3.5) or topographic conditions will not permit the construction of the exit channel bottom with a slope that falls within the required ranges, the need for a piped or concrete lined spillway is indicated.

Wherever there is a good vegetative cover in the spillway area and the topography (such as a natural saddle) is suitable, first consideration should be given to the use of a natural spillway.

Table 3.4 Minimum inlet width of spillway.

Spillway discharge (m³/s)	Depth of flow (m)					
	0.30	0.40	0.50	0.60	0.70	0.80
	Inlet width (m)					
1	4	4	4	4	4	4
2	7	5	4	4	4	4
3	11	7	5	4	4	4
4	15	10	7	4	4	4
5	19	12	8	6	5	4
6	23	15	10	7	6	4
7	27	17	12	9	7	5
8	31	20	14	10	8	6
9	35	22	16	12	9	7
10	39	25	18	13	10	8
12	47	30	21	16	12	10
14		35	25	19	14	11
16		40	28	21	17	13
18		45	32	24	19	15
20		50	36	27	21	17
22			39	30	23	19
24			43	32	25	20
26			47	35	28	22
28			50	38	30	24

Source: modified WAWA, 1991.
Notes:
i These widths are for well grassed spillways. Poorly grassed should be wider.
ii For practical purposes the minimum width has been taken as 4 m and the maximum 50 m.
iii The inlet width in the shaded area should be $^2/_3$ of the outlet width obtained from Table 3.5.

3.3.4 Chute spillways

Chute spillways on farm dams are not common because of their high cost, but they may be used where grassed earthen spillways are not possible. They are commonly built of concrete, but may be of sheet steel, bituminous concrete, or other erosion resistant material such as plastic sheeting. A natural rock surface, or a surface allowed to erode down to hard rock may also be regarded as a chute spillway.

The materials used in chute spillways are harder, and more expensive than earth. Therefore:

i   high velocities can be tolerated because of better erosion resistance;
ii   greater depth of flow and higher velocities are used to reduce the overall size; and
iii use of efficient hydraulic shapes to further reduce the size and cost becomes important.

Thus, instead of spreading the flow as widely as possible for grassed earth spillways, a narrow and deep chute is usually used. The inlet should occur over a relatively short distance in the direction of flow to achieve a high weir discharge coefficient.

Table 3.5 Minimum outlet width of spillway.

Spillway discharge (m³/s)	Return slope (m)									
	3%	4%	5%	6%	7%	8%	9%	10%	15%	20%
	Outlet width (m)									
1	6	6	6	6	6	6	6	6	6	6
2	6	6	6	6	6	6	6	6	7	9
3	6	6	6	7	8	8	9	10	14	18
4	6	6	8	9	10	11	13	14	19	23
5	6	8	10	11	13	14	16	17	23	29
6	8	10	12	14	16	17	19	20	28	35
7	9	12	14	16	18	20	22	24	33	41
8	11	13	16	19	21	23	25	27	37	47
9	12	15	18	21	24	26	29	31	42	
10	13	17	20	23	26	29	32	34	47	
12	16	20	24	28	31	35	38	41		
14	19	24	28	33	37	41	45	48		
16	22	27	33	38	42	46				
18	25	31	37	42	47					
20	27	34	41	47						
22	30	38	45							
24	33	41	49							
26	35	45								
28	39	48								

Source: modified WAWA, 1991.
Notes:
i   These widths are for well grassed spillways. Poorly grassed should be wider.
ii Lined spillways may be more economic in the range above 47 m.

Additional surcharge, and therefore freeboard on the embankment, will normally be economical to reduce the width of the inlet. The plan shape of the spillway will normally be wider at the inlet than further down the slope to take advantage of the accelerating flow velocities down the slope and thereby reducing the amount of concrete required.

Finally, because of the smooth, high-velocity flow down the chute, the water arrives at the bottom with high kinetic energy per metre width. This energy has to be dissipated in a structure specifically designed for the spillway if serious downstream erosion is to be avoided. However, this is not usually necessary for grassed spillways because most of the energy is dissipated on the way down the slope (SCA, 1979).

3.4   PIPELINES THROUGH EMBANKMENTS

Pipelines are often placed through or under embankments for any of the following purposes:

- to provide a gravity supply from the storage;
- as a suction pipe for a pumped supply (this always maintains a positive prime to the pump);

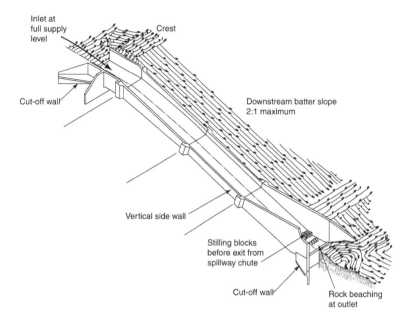

Figure 3.9 Chute spillway.

- to maintain flow downstream from the storage; and
- to bypass a significant flood-flow through a piped spillway.

Pipelines may be constructed using reinforced concrete pipes (rubber-ring joints), cement-lined cast iron pipes, galvanised or black steel pipes, high density polyethylene or PVC pipes.

Special care needs to be taken with joining of rubber-ring jointed pipes. An anchor block at the outlet valve is essential to avoid movement of the pipes. If concrete pipes are used the class of the pipe suitable for the depth of fill above must be used. PVC is liable to brittle fracture if impacted by construction equipment. All pipes should be pressure-tested to check welds, joints and seams on placing. In addition, a further test may be required after covering the pipe is completed, taking into account:

- the proposed full supply level;
- the superimposed load on the pipe;
- permit conditions if applicable; and
- siltation of the entry.

The pipe should be located underneath the embankment in a trench excavated in natural material. If required, the embankment core trench should be deepened where it crosses the outlet pipe to a depth of at least 0.6 m below the pipe. If outlet levels permit, the pipe should be located so that the tops of the core collars are at or below the natural surface to minimise the risk of damage by construction equipment. The

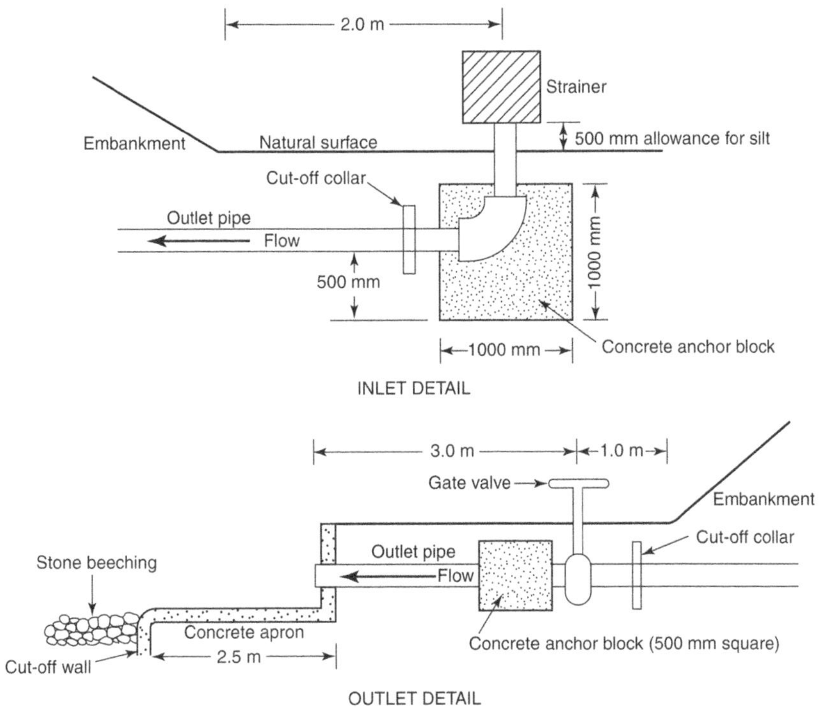

Figure 3.10 Inlet and outlet pipeline layout.

barrel of the pipe should be evenly and firmly bedded in the trench and should not be laid directly on rock or stony ground.

Cut-off collars are required to prevent seepage flows along the pipe. The pipe should always be laid in a trench, and the cut-off collars should extend down and sideways ($1 m^2$), well into natural ground or fully compacted material. An absolute minimum of two and preferably three cut-off collars should be provided.

The pipe should be positioned so that it is trenched into natural ground, and so that the inlet strainer extends into an excavated area, in order to avoid blockage by silt. If this is not possible, the use of a bent pipe with a vertical strainer may be necessary.

For small pipes (150 mm), heavy galvanised steel is usually used. For larger pipes, high-density polyethylene with concrete surround will probably be the most economic (in terms of installation) and will have the best life. Reinforced concrete with rubber ring joints can be used with care and adequate thrust and anchor blocks. Class–12 PVC (with concrete surround) can be used, but is not favoured. The chance of brittle fracture during construction is quite high. The type and class of pipe used must be selected on the basis of its suitability for the height of fill and bedding conditions.

Pipe smaller than 150 mm in diameter should not be used because of the danger of becoming clogged. The crest elevation of the emergency spillway should be located a distance above the invert or crest elevation of the trickle pipe inlet at least equal to the value of the minimum head required to provide full pipe flow.

3.4.1 Trickle pipes

A small pipe spillway is provided in many farm dams to protect the vegetative cover in the emergency spillway from prolonged saturation by continuous flow, spring flow, or flows that may continue for several days following a storm. This type of spillway is designed to discharge such a small percentage of the peak flow that it has no measurable effect on the emergency spillway design.

The design capacity of a trickle pipe should be adequate to discharge long-duration, continuous, or frequent flows without flow through the earth spillway.

3.4.2 Drop inlet structures

Generally a drop inlet structure consists of a vertical pipe (concrete or steel) connected to a horizontal pipe (fitted with cut-off collars) which passes through the bank and takes the flow away clear of the toe. These structures can be most useful on a storage with a small catchment area where:

- all the flood flows might be flood-routed in surcharge and all the discharge can go through a drop inlet;
- on a steep gully where a normal spillway is not practicable; or
- where the gully has an appreciable base flow.

Two precautions are taken to prevent unstable (full pipe) flow conditions from occurring:

i   There must be a significant gradient on the outlet pipe towards the downstream end so that any flow entering the pipe through its inlet orifice can flow away at less-than-full pipe flow. The entrance should be sharp, not smoothly transitioned, to retain the orifice effect.

ii The top of the outlet pipe must be ventilated immediately downstream of the drop pipe so that any air which is evacuated by mixing with the water in the outlet pipe is immediately replaced at atmospheric pressure. This vent pipe has to extend above maximum surcharge level of the storage.

It is also an advantage to have the drop pipe diameter about three or more times the diameter of the outlet pipe, so that weir flow is likely to be the discharge control under most surcharge conditions. This ventilated system will have less capacity than one of the same size allowed to flow full, but is a much more satisfactory and hydraulically stable system in practice. If greater capacity is required, a larger sized pipe should be used.

Calculations of discharge and head through drop inlet spillways have to be done as a three stage process checking:

1   the head-discharge over the inlet weir;
2   the head discharge at the orifice entry of the outlet pipe; and
3   the flow conditions in the outlet pipe, to ensure part-full flow.

The weir is essentially a pipe spillway with a significant flow, that is the flow is taken into account when determining the emergency spillway.

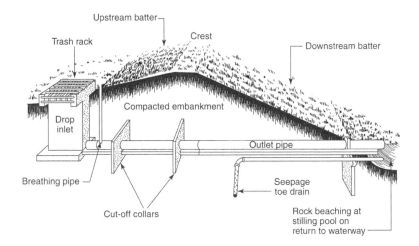

Figure 3.11 Drop inlet structures (*Source*: modified SCS USDA, 1969).

Table 3.6 Separation distance of cut-off collars.

Length of pipe (m)	<20	20	25	30	40	60
Number of collars	3	4	5	6	8	12

Note:
An absolute minimum of 3 cut-offs located upstream should be used. Cut-off collars may be constructed of concrete, mild steel plate or PVC.

3.4.3 Cut-off collars

The object of cut-off collars is to prevent seepage along the barrel of the pipe (see Table 3.6).

3.5 EARTH AND WATER COMPUTATIONS

3.5.1 Embankment material

The estimate of the volume of material required should include the embankment, allowance for settlement (see Section 5.7), backfill core trench (see Section 5.4.5), stripping topsoil (see Section 5.4.4), backfill gullies and any other banks that the contractor is required to construct, for example, diversion banks.

Volume estimates for dams are made on the basis of cubic metres (m^3) of earth fill in place, that is, compacted volume and not excavated volume or loose (transported) volume. The symbols used for the calculation of all earthwork computations in this Section are shown in Figure 3.13.

Formula for calculating the volume of earthwork:
This formula can be stated as:

$$V = 0.5P \times (A_1 + A_2) \tag{1}$$

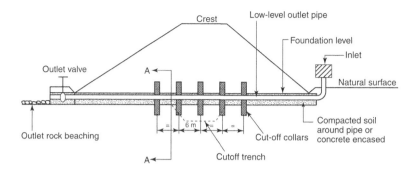

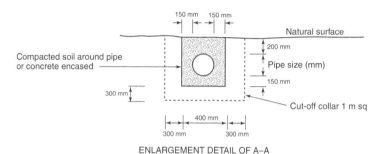

ENLARGEMENT DETAIL OF A–A

Figure 3.12 Pipeline through embankment with cut-off collars.

where:

V = volume in cubed metres (m³)

A_1 & A_2 = cross-sectional area at different points along an embankment (m²)

P = perpendicular distance between end areas A_1 & A_2 (m).

For more than two cross-sections then:

$$V = \text{olume}(m^3) = P \times (0.5A_1 + A_2 + A_3. + \ldots A_{n-1} + 0.5A_n) \tag{2}$$

where

n = number of cross-sections A_1, A_2, A_3…A_n involved.

Table 3.7 can be used to give the cubic metres per metre length for A_1, A_2, A_3 & A_4 for the different crest width, fill height and batter slopes of an embankment.

Example

An estimate of the volume (m³) of material for an embankment in a gully is required. The following information has been provided in Figure 3.13.

Cw = Embankment length = 25 m

T = Crest width = 3 m

H = Fill heights (m): H_1 = 4.00, H_2 = 4.5, H_3 = 4.25 and H_4 = 3.75

Batter slopes: a = 3: 1 (U/S), b = 2:1 (D/S)

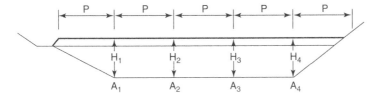

Cross-sectional view A-A of embankment

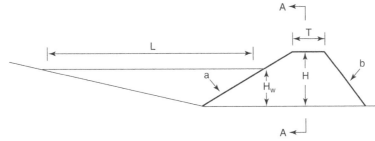

Longitudinal view – embankment and storage

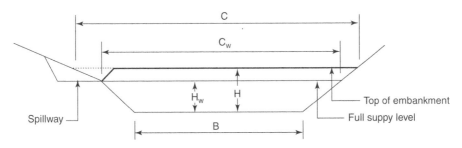

Embankment symbols used in formula

Figure 3.13 Computational symbols (*Source:* SCA, 1983).

As the 4 cross-sectional areas are evenly spaced P = 5 m; otherwise P the perpendicular distance between end areas (A_l, A_2, A_3 & A_4) would need to be calculated separately for each segment.

By using information provided, that is, crest width (T), fill height (H), batter slope ('a' U/S & 'b' D/S), and Table 3.7 – fill required for earth dams, the volume (cubic metres per metre – m^3/m) can be obtained between each cross-sectional area for use in formula (2):

Areas at each section A_1, A_2, A_3 and A_4 from Table 3.7 are:

$A_1 = 52.0 (m^3/m)$,

$A_2 = 64.1 (m^3/m)$,

$A_3 = 57.9 (m^3/m)$,

$A_4 = 46.4 (m^3/m)$

Substitute this data into formula (2) gives:

$$V = P \times (0.5A_1 + A_2 + A_3 + 0.5A_4)$$
$$= 5 \times (0.5 \times 52.0 + 64.1 + 57.9 + 0.5 \times 46.4)$$
$$= 5 \times 171.2$$
$$= 856\,\text{m}^3$$

3.5.2   Floor slope

All methods of calculating volumes assume a horizontal valley floor. This is generally not the case. For floor slopes up to 5 per cent and banks of up to 6 m high the difference in volume is not significant (less than 4 per cent).

3.5.3   Area beneath embankment

It is sometimes necessary to know the volume of stripping required from beneath the embankment. This volume can be calculated by the use of the formula:

Volume = depth of stripping x area beneath banks

$$V = S \times P(\underline{CT} + (a + b)\Sigma H) \qquad (3)$$
$$(P \qquad\qquad)$$

$V =$ volume (m^3)
$S =$ epth of stripped topsoil (m)
$P =$ distance between sections (m)
$C =$ length of embankment (m)
$T =$ crest width (m)
$\Sigma H =$ sum of heights (m) at all sections
$a + b =$ upstream + downstream slopes
i.e. 3H :1V $= a = 3$, 2H :1V $= b = 2$

Example

Use the data from the example in Section 3.5.1 to find the volume (m^3) of material to be stripped.

$$S = 0.10\,\text{m}$$
$$P = 5\,\text{m}$$
$$C = 25\,\text{m}$$
$$T = 3\,\text{m}$$
$$\Sigma H = 4.00 + 4.50 + 4.25 + 3.75 = 16.5\,\text{m}$$
$$a + b = 3H : 1V \text{ is } 3, 2H : 1V \text{ is } 2 = 5$$

Substitute data into formula (3) gives:

$$V = S \times P \left(\frac{CT}{P} + (a+b)\Sigma H \right)$$

$$V = \frac{0.1 \times 5}{5} \underline{(25 \times 3} + 5 \times 16.5)$$

$$= 48.7 m^3$$

3.5.4 Excavated tanks

Volumes of excavated tanks should be calculated by the prismoidal formula which is:-

$$V = \frac{d(A_1 + 4A_m + A_2)}{6} \qquad (4)$$

V = volume (m^3)

d = depth between A_1 and A_2 (m)

A_1 = top natural surface area (m)

A_m = mid area (m^2)

A_2 = bottom excavated area (m)

3.5.5 Water storage capacity computations

Storage volume for small gully dams can be approximated by the use of either formula (5) or (6):-

$$V = \frac{(H_w \times L)(2B + C_w)}{6 \times 1000} \qquad (5)$$

where the symbols are shown in Figure. 3.13 and

V = capacity in megalitres (ML)

H_w = depth of water at Full Supply Level at embankment (m)

L = distance from embankment to tailwater at Full Supply Level (m)

B = length of embankment at bed level (m)

C_w = length of embankment at Full Supply Level (m)

Example

Use the data from the example in Section 3.5.1 to find the volume (m^3) of water stored.

H_w = 3.5 m (includes 1.00 m freeboard)

L = 50 m

B = 15 m

C_w = 25 m

Table 3.7 Fill required for earth dams.

	Crest width (T)											
	2.5 (m)				3 (m)				3.5 (m)			
	Batter slopes (a = Upstream : b Downstream, i.e. Horizontal :Vertical)											
(a) U/S	3½:1	3:1	3:1	3:1	3½:1	3:1	3:1	3:1	3½:1	3:1	3:1	3:1
(b) D/S	3:1	3:1	2½:1	2:1	3:1	3:1	2½:1	2:1	3:1	3:1	2½:1	2:1

Fill
height (m) Volume of fill required cubic metres per metre (m³ per m)

Fill height (m)												
0.25	0.83	0.82	0.80	0.79	0.95	0.94	0.92	0.90	1.08	1.06	1.05	1.03
0.50	2.1	2.0	1.9	1.9	2.3	2.3	2.2	2.1	2.6	2.5	2.4	2.4
0.75	3.7	3.6	3.4	3.2	4.1	3.9	3.8	3.7	4.5	4.3	4.2	4.1
1.00	5.8	5.5	5.3	5.0	6.3	6.0	5.8	5.5	6.8	6.5	6.3	6.0
1.25	8.2	7.8	7.4	7.0	8.8	8.4	8.0	7.6	9.5	9.1	8.7	8.3
1.50	11.1	10.5	9.9	9.4	11.8	11.3	10.7	10.1	12.6	12.0	11.4	10.9
1.75	14.3	13.6	12.8	11.9	15.2	14.4	13.7	13.0	16.1	15.3	14.5	13.7
2.00	18.0	17.0	16.0	15.0	19.0	18.0	17.0	16.0	20.0	19.0	18.0	17.0
2.25	22.1	20.8	19.5	18.2	23.2	21.9	20.7	19.5	24.3	23.1	21.8	20.5
2.50	26.6	25.0	23.4	21.9	27.8	26.3	24.7	23.1	29.1	27.5	25.9	24.4
2.75	31.5	29.6	27.7	25.8	32.8	30.9	29.0	27.1	34.2	32.3	30.4	28.5
3.00	36.8	34.5	32.2	30.0	38.3	36.0	33.8	31.5	39.8	37.5	35.3	33.0
3.25	42.5	39.8	37.2	34.2	44.1	41.4	38.8	36.2	45.7	43.1	40.4	37.7
3.50	48.6	45.5	42.4	39.4	50.3	47.3	44.2	41.1	52.1	49.0	45.9	42.9
3.75	55.1	51.6	48.0	44.4	57.0	53.4	49.9	46.4	58.8	55.3	51.8	48.3
4.00	62.0	58.0	54.0	50.0	64.0	60.0	56.0	52.0	66.0	62.0	58.0	54.0
4.25	69.3	64.8	60.3	55.8	71.5	66.9	62.4	57.9	73.6	69.1	64.5	60.0
4.50	77.1	72.0	66.9	61.9	79.3	74.3	69.2	64.1	81.6	76.5	71.4	66.4
4.75	85.2	79.6	73.9	68.3	87.6	81.9	76.3	70.7	90.0	84.3	78.7	73.0
5.00	93.8	87.5	81.3	75.0	96.3	90.0	83.8	77.5	98.8	92.5	86.3	80.0
5.25					105.3	98.4	91.5	84.6	108.0	101.1	94.2	87.3
5.50					114.8	107.3	99.7	92.1	117.6	110.0	102.4	94.5
5.75					124.7	116.4	108.2	100.0	127.6	119.3	111.0	102.7
6.00					135.0	126.0	117.0	108.0	138.0	129.0	120.0	111.0
6.25					145.7	135.9			148.8	139.1		
6.50					156.8	146.3			160.1	145.9		
6.75					168.3	156.9			171.7	160.3		
7.00					180.3	168.0			183.8	171.5		

Source: SCA, 1983.

Substitute values into formula (6):

$$\text{Volume} = \frac{(H_w \times L)(2B + C_w)}{6 \times 1000}$$

$$= \frac{(3.5 \times 50)(2 \times 15 + 25)}{6 \times 1000}$$

$$= 1.6\,\text{ML}$$

$$V = \frac{K \times A \times H_w}{1000} \tag{6}$$

where V = capacity in megalitres (ML) A = surface area of water stored at full supply
level (m^2)

H_w = depth of water at embankment (m)

K = constant with default value 0.4. The value of K increases as the side slopes of
land become steeper, reaching K = 1 for vertical sides.

Example

$$K = 0.4$$

$$A = 50 \times 25 = 1250\,m$$

$$H_w = 3.5\,m$$

Substitute data into formula (7):

$$= \frac{0.4 \times 1250 \times 3.5}{1000}$$

$$= 1.75\,ML$$

When an accurate assessment of the storage volume is required (e.g. irrigation dam), a contour plan should be produced by grid survey or tacheometry.

Storage volume can then be calculated by the use of a planimeter and end area formula.

All the above measurements can be made on site with a level or theodolite. This can be achieved either in the form of a cross-section survey at the centre line of the proposed dam or by a contour survey. The latter approach is more accurate and time consuming, but also more useful where comparison of similar sites is being undertaken.

Using the formula, approximate figures for the capacity at various heights of dam can be derived. The capacity estimated in this way is accurate to within 20 per cent, but it must be revised by a more detailed survey when the site has been approved for possible construction. An approximate formula considers the water volume to be an inverted triangular pyramid with surface area of water $= (^{LT}/_2)$ and depth $= (^D/_3)$. With experience, one is able to judge fairly accurately how an individual valley will compare with such an idealised picture and therefore to adjust the resulting conclusions.

If all the excavation is from below the full supply level contour the volume held in storage is the volume calculated from formula above plus excavated volume minus $^1/_2$ bank volume. This is approximately the same as storage volume plus volume calculated above plus $^1/_2$ excavated volume.

3.5.6 Storage excavation ratio

The storage excavation ratio is a measure of the unit cost of the stored water. It is the ratio between the total storage and total earth moved. See Section 1.4 for different types of farm storages.

3.6 ESTIMATE OF COSTS

The term 'estimate of costs' means an estimate of the contract cost of construction of physical works and the purchase of items of equipment. Cost estimates are an essential part of the process of decision making in farm projects. They are used by:

- the designer in deciding which option is the best design solution;
- the owner/client in deciding whether the proposed dam is likely to be profitable; and
- the contractor, who prepares estimates on the basis of how much to charge for the work. A contractor has to estimate very accurately to stay in business.

In some circumstances cost estimates may also be used by finance organisations to decide on the approval of money for the scheme. They have also been used as a convenient method of calculating charges for planning work, particularly subsidised charges made by Government services.

The accuracy of the estimate is often determined by the proposed use. Very rough approximations may be quite suitable for comparing different solutions to a design problem. In fact, if very accurate estimates are needed to choose between options, then they are so close in cost that the choice may be made on other factors. Similarly, very close estimates may not be warranted for decision making on profitability, especially when the part of the project being estimated is not a major part of the total cost, as is often the case with farm dams. Generally, estimates to arrange finance for a project should be fairly accurate and will usually be finalised only after agreement is reached on details of the design. The most accurate estimates are those prepared for contract purposes by the contractor to determine a quotation figure and by the designer for comparison with contract prices. The contractor has to be accurate – too low and a loss is incurred, too high and the contractor does not get the project. Both scenarios relate to the viability of the business. The design estimate needs to be accurate, and not disclosed to contractors, so that quotation prices may be compared on an objective basis. Quotations well below a design estimate should be examined very closely to sort out, in advance, problems that are likely to arise during construction.

3.6.1 Economics

The value of land occupied by the work may be important and may influence the design. It will usually be necessary to include land value in farm dam estimates only when comparisons are being made between different design options.

A farmer's decision to build a small and deep tank when a large shallow tank would have much cheaper earthworks will usually be determined by the value of the additional land which is occupied or disrupted by the lower cost storage.

Table 3.8 Schedule of rates contract.

Item	Description	Rate	Quantity	Amount
1	Site Engineers Fees	10%	....	....
2	Soil Testing	Sum	....	....
3	Land Survey	Sum	....	....
4	Licensing Authorities Fees	Sum	....	....
5	Vegetation & Topsoil Stripping	$-\$/m^3$	....	....
6	Excavate Core Trench	$-\$/m^3$	....	....
7	Earthworks–Bank/Core Trench	$-\$/m$	....	....
8	Outlet Structures	$-\$/m$	....	....
9	Embankment – Construct	$-\$/m^3$	....	....
10	Levee Banks	$-\$/m^3$	....	....
11	Bywash – Spillway Construction	$-\$/m^3$	....	....
12	Replacement of Topsoil on Bank	$-\$/m^3$	....	....
13	Gravel on Access Track	$-\$/m^3$	....	....
14	Seed and Plants	$-\$/m^2$	....	....
15	Fence Out Areas	Sum	....	....
16	Gauge Boards	Sum	....	....
17	Other(s)	Sum	....	....

Sum Total
GST@.........
Extras@x..%
Total$......

Farmers may seek to save money by building dams themselves, either with their own machinery or with hired plant, particularly if the farmer has time available which would otherwise not be used to earn more than the savings on a contract construction price. This has the advantage that the dam builder has a direct interest in achieving a good result. However, these potential savings should not be deducted from the cost estimate.

The estimate of costs should be based on rates which would reasonably be expected to cover construction and materials. Should the client be able to obtain a quotation at a cost lower than the design estimate, there is no reason why it should not be accepted, provided the contractor has suitable equipment and agrees to carry out the work in accordance with the plans and specifications.

For many purposes the future annual cost of a dam is important. This cost determines whether the project is profitable (compared with the annual value of increased production). It is also needed to combine with other annual costs to determine the best project design. For example, a dam built at low capital cost, but which had high annual cost for repairs and maintenance, might well cost more than one built to a higher standard and initial cost with minimal maintenance cost. The objective is an optimal combination of annual costs of capital and recurring items. Comparison of all options should be considered before proceeding with a choice of project.

Contractors generally work on an hourly rate of hire for each piece of construction plant. The hire charge is negotiated between the landowner and contractor before work commences.

Other forms of contract can be used:

i Schedule of rates contract

When using this method of contract the actual measurement of each item becomes very important.

ii Lump sum contract

The contractor estimates the cost of construction making allowances for any unknown factors that may affect the cost. This form of contract, while common in structural engineering jobs, is not used to any great extent for earthmoving works because of the greater number of unknown factors, including water and the inconsistency in earth materials (SCA, 1983).

A mixture of the above forms of contract may sometimes be used, for example, an hourly rate for the core and a lump sum for the embankment.

3.6.2 Dam quality pays over time

Dam quality does not cost – it saves. There are points to consider when planning to build a dam in relation to cost. In general, you get what you pay for, and good quality work pays for itself in the long run (Lewis, 1995).

In any kind of business transaction it is common sense to expect good value for money. The same argument applies to employing a contractor to build a dam on your farm. The lowest quote may turn out to be expensive in the end and the person whose quote is much cheaper than another's may only do half the job. Whoever you get to design and build your dam, make sure you have considered the points provided in Section 5.2. Building a dam is costly, and it is well worth your time and money to make sure you get a top quality product.

Documentation

An engineering plan for the dam should be developed using information collated during the design process.

This plan should include:

- all pertinent elevations;
- dimensions;
- extent of the cut-off trench;
- other areas requiring backfill;
- location;
- dimensions of the trickle pipe and other planned outlet structures; and
- all other factors crucial to the construction of the dam.

It should also take into consideration a schedule of materials documenting the quantity and type of all the necessary construction materials. To ensure that the quality of construction meets requirements, both the landowner and contractor should receive copies of the plans and specifications to assist them in understanding the demands.

The Australian National Council of Large Dams (ANCOLD) and most Australian State water authorities have developed uniform engineering standards and construction specifications for dams.

4.1 COLLATION OF PLANS AND SPECIFICATION

Engineering plans are prepared methodically through a process of collection, recording and analysis of the facts and data to obtain a satisfactory outcome to a project. The final product is generally a written report, with diagrams and tables, outlining the class, scope and quality of work to be accomplished.

All of the material presented in this section is to be used as a guide in analysing site conditions and preparing engineering plans. It is therefore critical that designers collect sufficient data to ensure the elimination of inherent problems after they have been identified. However, it is equally important to observe an orderly approach in the collation of site data, the consideration of alternate possibilities, and the use of accepted procedures for the development of designs and plans.

Certain basic procedures should be taken in examining the project and in preparing relevant engineering plans, regardless of the size and complexity of the work. Indeed, the same process of thought and action applies even when the job involves a small dam.

The basic steps are shown below.

1   Identify the project and its scope.
2   Investigate site conditions.
3   Collect all relevant design data.
4   Examine and collate data.
5   Design plan based on the above conditions.
6   Prepare plans and specifications for construction.
7   Review the plan and seek approval from responsible authorities.

4.2   COLLECTING BASIC DESIGN DATA

Subsequent items represent data that may be applicable in the development of an engineering plan. Not all of these items may apply in every situation, and furthermore, additional investigations may be required for more complex projects. The designer should carefully analyse the available data in order to ascertain what additional information is required.

4.2.1   Catchment map

All catchment information relevant to the analysis and design of the proposed works should be assembled and recorded on a prepared form or map. The amount of detail needed will depend on the complexity of the structure. Catchment information includes the area of the contributing catchment, physical characteristics of the catchment, and the location of the proposed works within the catchment. Such information may include some or all of the following:

1   average slope of various lengths of the main waterway;
2   average slope of the land in various parts of the catchment (generally this can be obtained from a 1:25 000 topographical map of the area);
3   categorisation of land use, that is, irrigation, grassland and woodland;
4   the main soil types or groups within the catchment;
5   the location of the proposed dam wall and submerged land, i.e. storage;
6   subdivisional information and title description; and
7   proprietary interest on large catchments.

In order to prepare a catchment map, any combination of the following methods may be used:

• mark information on an aerial photograph;
• relay data from aerial photograph onto a work sheet;
• proportionally enlarge data from aerial photograph onto cross-sections; and
• draw information on a 1:25 000 topographic map.

4.2.2 Location (topographical) map

A location map is used for planning details and for outlining the project. It is frequently referred to as a plan view of the proposed works. It is normally drawn to a scale larger than the catchment map, but may be combined with a catchment map on some of the smaller projects, especially on small drainage work. A location map should show as many of the following items as required in the design, construction, and future maintenance of the project.

- Position of survey centre line, or other survey lines, connected to land fixtures.
- Location and plan view of all features (including horizontal dimensions) of proposed works with reference to survey lines.
- Location and elevation of bench marks at Australian Height Datum (AHD).
- Location of existing waterways (with names) or other features when these affect design.
- Location of soil testing holes to a Geographic Information System (GIS).
- Surface or sub-surface water elevations at the time of investigation.
- Location of existing fences, property lines, buildings, roads, culverts, bridges, springs, wells, borrow pits.
- Contour lines to Australian Height Datum (AHD).

4.2.3 Profiles and cross-sections

Defining the detailed shape and elevation of the existing terrain of the site is often necessary in the preparation of engineering plans. Generally, field data provide baseline information to compile plan views, profiles, longitudinal and cross-sections. The detail and accuracy of the survey should be in line with the complexity of the site and the structure design.

4.2.4 Soils

The engineering characteristics and land capabilities of the soil are important in determining the suitability of a dam site. Records of soil holes, cut trenches or other investigations are used to determine:

- the ability of the foundation materials to support the structure adequately;
- the ability of a storage site to retain water;
- suitability of the materials for embankment;
- permeability of soils for effective drainage systems; and
- depth to rock, groundwater, or other conditions that may affect the structure.

The depth, method, and scope of the soil and foundation investigations will vary according to the size of the structure and the hazard of the site. Generally, adequate investigation can be made with the use of a hand auger. When inspection of undisturbed soils is required, the use of heavier equipment, such as backhoes and bulldozers, has proven to be efficient and economical. All sites at which investigations are made should be numbered and plotted on the engineering plan, and the findings recorded.

Most storage dams control, store, or provide discharge capacity for a certain volume or flow of water. The expected safety and efficiency of the dam is related to the accurate determination of the design run-off volume, or peak discharge of the contributing drainage area or other source of supply.

Certain site data and capacity information are required to evaluate the hydraulic requirements of the dam. Items that often require consideration are:

- the alignment and slope of waterway;
- grades and cross-sections;
- the critical elevations;
- upstream and downstream capacities and controls;
- the stream-bed conditions, that is, bed load and deposition patterns that affect velocity and erodability;
- rate of release from controlled supply;
- the design flow from hydrologic data; and
- earthquake and structural fault lines.

4.3 ASSEMBLY OF DATA

After collecting all relevant site data, it is important that all the facts and information be arranged and recorded in an orderly manner. Survey notes should be used in the assessment and in plotting plans in accordance with standard procedures. Time can be used more efficiently by preparing the engineering plans to scale and in detail, as the field data is obtained. This will generally suffice for final design and preparation of plans for the project.

For many projects the use of approved, standard work sheets can (and should) be used to record all of the necessary data. This information can then be used later in the project, if problems arise.

4.3.1 Analysis of data

It is important to consider the relationship of the proposed solution to the overall use of the land, when reviewing the collected data.

- Detailed review of the data is necessary to determine:
- the type or series of measures required;
- the limitations in location and size of facility imposed by the site; and
- the design procedures and criteria that apply to the proposed structures or practices.

Throughout the process of analysis, consideration needs to be given to alternative methods and construction materials. A small, additional outlay in installation cost may prove a saving in future maintenance and operation.

4.3.2 Design

Design procedures include:

1   rechecking or amending the estimated catchment flow regime;

2   hydraulic calculations to provide information on control, capacity, and safety;
3   structure design (see Section 3);
4   location, dimensions and elevations of important parts of the dam wall and its accessories;
5   estimate of material quantities (see Section 3.5);
6   estimate of construction costs (see Section 3.6); and
7   specifications for materials and construction (see Appendix 2).

It is important that design calculations be documented in an orderly manner and checked for accuracy.

4.4 CONSTRUCTION DOCUMENTS AND DRAWINGS

The preparation of the construction plans or drawings provides a detailed record of the design and structural requirements of the dam. This allows a contractor who is unfamiliar with the project to lay out and ensure that the work is constructed as designed. Construction drawings for complex dams usually involve drafting the topographic, cross-section and profile data collected in the investigation stage. Additional layout and detail drawings, as required, may be prepared as the design calculations are completed.

For many on-farm measures, the preparation of the construction drawings may be simplified by using approved standard pre-design layouts. The standard drawings require careful review to see that:

•   the plan will fit the site so that the structure will function properly; and
•   all required dimensions, elevations and modifications are complete.

A common error in the preparation of the construction drawings is the omission of required details, sections, and dimensions. All plans should be carefully reviewed for completeness and accuracy.

4.4.1 Specifications

In addition to the detailed construction drawings, the construction plan often requires written specifications to clarify how the work will be done, the quality of workmanship, and methods of testing (see Appendix 2). Another important component of the specifications is the required quality of the manufactured materials that will be used in the work.

For small projects, the material and construction specifications may be documented in the form of notes on the drawings. For larger projects, the preparation of a separate specification document, or the use of Standards Association of Australia (SAA) or ANCOLD or State guidelines are more practical. In all cases, readability of drawings is vital to the success of the project.

4.4.2   Checklist

The amount of detail required in the construction plans will vary with the type and size of the project. However, all projects regardless of size should be adequately planned. The following list may be useful in checking the adequacy of the drawings and specifications.

1   Can the farm be located from the plans?
2   Is the project site clearly shown?
3   Can the survey lines be relocated and the job pegged for construction as designed?
4   Are all dimensions and construction details clearly shown?
5   Are material and construction specifications complete for all parts of the work?
6   Are material quantities shown?
7   Has the title block been completely filled, including the date and who designed, drafted, amended and approved the work?
8   Has the cost estimate been prepared?
9   Have licence/permit documents been lodged with the responsible authorities?

4.5   FINAL REVIEW AND APPROVAL

Before the construction plans are delivered to the contractor, the design and construction plans should be reviewed by an independent designer.

When required, plans should be submitted to the appropriate responsible authorities for approval. The construction plans should also be properly signed and dated by all parties involved in their preparation and approval.

4.5.1   Records

A complete copy of survey notes, useful basic data, soils logs, design calculations and other relevant data, including a copy of the plans and specifications, should be assembled and filed in an orderly manner.

All dams or practices, regardless of type or kind of material, will require maintenance. Changes or additions made during construction should be recorded in coloured pen on the office copy of the plans. These 'as constructed' plans are often useful when making maintenance recommendations to the contractor. They are also useful for structural design improvement and for the evaluation of hydraulic performance. Moreover, complete 'as constructed' records may be valuable in case of a legal dispute (see Section 10).

Construction

Despite the apparent super-efficiency of modern equipment, a high proportion of dams built today do not seem to have the long life of those small dams constructed in the old days with a horse and scoop, followed by a driven flock of sheep to give compaction. Although bulldozers, scrapers and tractors are heavy machines, they are designed to create as little pressure as possible and are usually used when the soil is at its driest – a time when soil moisture is well below the optimum for good compaction. The modern answer for the compaction process is a wheeled tractor or sheepsfoot roller, particularly where troublesome soils are encountered, and a crawler tractor to carry out the excavation.

5.1 APPROVAL FOR DAM BUILDING

Under the provisions of different State Legislation in Australia, people wishing to build dams, including weirs, must have the approval of the responsible authorities before commencing any works (see Section 10).

It is important, first, to contact the responsible authorities to ascertain whether the proposed dam would need to conform with design, planning and licensing requirements. In some States of Australia (see Section 10, Table 10.1), certain procedures must be followed when applying for the issue of a licence to construct. On satisfactory completion of the works a licence containing guidelines and controls is issued. This is necessary to safeguard water supplies, the stream environment and existing water users' requirements.

Upon receipt of the information needed, the responsible authorities will make an evaluation, taking into account the water available, environmental issues, planning and any possible detrimental impacts on existing diverters. Submission of this information will not preclude additional information being requested when it is considered necessary.

5.1.1 Details that may need to be submitted

In order to assess an application and obtain authorisation to commence construction of a dam, an engineer/contractor/landowner may need to submit details to the responsible authorities.

Dams constructed in Victoria on waterways defined under the *Water Act* (1989) need to meet the following licence conditions:

- approvals in regard to the location of a storage dam from other relevant authorities such as the Environment Protection Authority (EPA), Catchment Management Authority (CMA) or Rural Water Authority (RWA);
- a planning permit from the municipality in which the proposed private dam is to be operated;
- an occupancy licence, if any part of the proposed works is to be located on Crown land, or adherence to conditions may be set if the dam is to be within a proclaimed water supply catchment; and
- approval for any instream works from the river management authority if such an authority exists.

Other requirements may relate to the storage of water:

- small dams may only be filled during the months May to October inclusive providing sufficient flows are available. From November to April inclusive, all incoming flows must be passed downstream through the low level outlet pipe to which a gate valve has been connected;
- a staff gauge in the storage to measure the water depth and 'V' notch weir upstream of storage is installed to allow water inflow fluctuations to be observed;
- maintenance is to be carried out to ensure continuing compliance with specified dimensions and levels and the safety of the structure; and
- appropriate written dam-sharing agreements, and approval with reference to potential flood risks on neighbours' properties, should be obtained before commencing any works on a dam.

Upon satisfactory completion of the storage construction, and payment of the relevant charges, the Water Authority will then authorise an annual licence. If the water is to be used for irrigation, then 1 ha of crop or pasture for each 3 ML of storage capacity may be used, depending on locality and the type of crop.

In New South Wales, the new *Farm Dams Policy* (1999) allows the use of 10 per cent of the mean annual run-off as a harvestable right. When water requirements exceed this amount, it is necessary to apply for a small dam licence (see also Section 10). The process of determining the harvestable right for a property, and applying for a licence, is managed by the New South Wales Department of Land and Water Conservation (DLWC). Supporting documentation is available from the DLWC website.

5.1.2 Referral dams

Many States in Australia have laws that require permits or licences to construct dams for water storage for any intended use. The engineer/contractor should know the requirements in their State and comply with them in planning, design and layout of dams. Referral dams, sometimes called prescribed dams in New South Wales, also come under the provision of most State legislation, and are defined as being:

- 5 m or more high with a capacity of at least 50 Megalitres, or
- 10 m or more high with a capacity of at least 20 Megalites. These dams will require additional information to be supplied. Further information on these requirements can be obtained by contacting one of the responsible authorities in that State.

The following requirements tend to apply to most States of Australia for dams in excess of wall height and/or storage capacity stated above, and, in certain circumstances, to dams of lesser size:

- they must be designed by a professional engineer who is experienced in the field of small dam engineering;
- certified engineering designs, plans, computations and specifications must be submitted to the responsible authorities for consideration before any works will be allowed to commence; and
- the construction of the works must be supervised by the designer or a similarly qualified professional engineer, who will certify that such works have been completed to design specifications and provide as-constructed details.

5.2 SELECTING YOUR DAM BUILDERS

Common causes for dam failure are listed in Section 6. Your contractor should encourage you to hire an experienced small dams engineer to plan your dam before the contractor starts moving any earth. The contractor will do a much better job for you if they have been provided with a plan. Do not jump to the conclusion that the contractor must be getting a commission from the engineer. A plan is vital to the success of the whole project, because it ensures that the money spent on the rest of the project is not wasted. Contractors who say they don't need a design plan are your worst enemies! Beware of the person who says, 'I can do it all and save you money'. This kind of 'bargain' can end up being very costly.

5.2.1 Selecting an engineer

Before building any large dam, talk to a consultant engineer experienced in small dam work. Furthermore, consult the engineer during the design and construction phases, in the event of problems or uncertainties with the dam contractor. Many small dams fail because of poor planning, poor investigation and design, unsatisfactory siting, faulty construction practices or lack of maintenance after construction. The larger the dam, the more critical these factors become. Many of the failures result in total loss of the dam. In those cases where the damage can be corrected, repair costs can be more expensive than the original cost of building the dam.

The engineer must consider a range of issues when designing a dam. The issues crucial to the dam's safety include:

- a good understanding of local geology – and whether soils in the vicinity of the dam are suitable for storing large volumes of water;

- knowledge of the properties of the foundation material beneath the dam – whether it will support the load without excessive deformation, and control seepage to within acceptable levels;
- understanding the materials from which the dam will be built – whether they have adequate strength, durability and low permeability, and from where they can be most economically obtained;
- a knowledge and understanding of spillway design requirements;
- an appreciation of local meteorological and hydrological conditions, in particular for determining spillway capacity. Too often the importance of an adequate spillway is not reflected in the initial planning. In many cases, determination of spillway capacity is simply based on normal winter flows of the nearby creek or stream. However, 'normal' flows can be exceeded many times during or after heavy rain, and the safety of the dam embankment depends entirely on the ability of the spillway to operate under peak storm flow conditions;
- a sound knowledge of relevant dam design principles and methods, to ensure safety, economy and durability. Experience in good practice covers such areas as foundation cut-offs, internal and external drainage, outlet pipes, trickle pipes and erosion protection; and
- a broad knowledge of other factors that may be relevant in particular situations, including knowledge of legal obligations under relevant State legislation and awareness of potential problems such as siltation, wave action, other forms of erosion, and pollution.

5.2.2 Details that an engineer can provide

The designer/engineer will need to supply the responsible authorities with a brief report outlining the design of the storage to obtain authorisation to commence construction of a dam. The report must contain the details shown below.

- Location of the proposed dam using the following parameters – Australian Map Grid co-ordinates, Australian Height Datum, Land Title description and distance from major landmarks or roads. It is desirable that other improvement sites are located within the proper section and subdivision as accurately as possible from existing map or field data.
- A brief description of the catchment area, its soil type, vegetation and topography. This may also contain legal land descriptions and the dam shown on the catchment or location map.
- Description of foundations and foundation material of the dam site, including the downstream area as far as the point of return of the spillway channel to the waterway.
- The nature and properties of the proposed embankment material (laboratory testing of material to be carried out by a registered company experienced in this field).
- An indication of the embankment design methods and cases, and the factors of safety adopted (including earthquake prone areas).
- Calculations of the design storm inflow (for at least a 1 per cent chance in 100 years of a recurrence interval storm) or such longer period as may have been specified.
- Calculation of the earth or culvert spillway size to pass the design outflow, including the downstream channel as far as the point of return to the waterway (earth

spillways are usually 1 m deep with a freeboard allowance of 500 mm under design flood conditions) and an assessment of possible erosion effects under design conditions.
- Plans of the proposed embankment showing full details and dimensions, including spillway, capacity of storage, compensation pipe and trickle pipe, berms, beaching, crossing culverts.
- A brief outline of housing, utilities or other development downstream of the storage.
- Details of the proposed method of construction and equipment to be used, the degree of supervision to be provided, and the names of the supervising engineers.
- When the dam has been constructed, the engineer will be required to certify that such works have been completed to design and to provide 'as constructed' details.
- On request the engineer may provide a maintenance and safety surveillance program for ongoing use.

5.3 HOW TO BUILD A DAM

Most of the element of risk can also be eliminated from the building of a dam by careful attention to several basic requirements. The builder should ensure that:

- the site selected is the most economical, that is, a natural depression is usually best because it considerably reduces the amount of excavation;
- the storage has low seepage losses and the soils present are capable of holding water – clay is usually the best (sometimes a sealer can be used with sandy clays, but this increases costs significantly);
- data relating to dam capacity, that is, the limitations of the drainage area are available;
- the foundations are watertight and capable of supporting the dam;
- there is sufficient dam-building material; and
- 75 per cent of the drainage area is permanent pasture or woodland to prevent excessive silting.

The cost and effort expended on making sure that these conditions are met will, in the long run, save the builder both time and money.

5.4 STEPS IN CONSTRUCTING A DAM

Building a good dam embankment is rather like making a layer cake; it is a methodical process requiring exactly the right ingredients and careful attention to building each layer. However, unlike a cake, the aim of building a dam wall is to force air out from between soil pores, making the soil denser and less permeable. The following steps should be considered before constructing an embankment, as listed in Appendix 2.

5.4.1 Setting out

Setting out or pegging out transmits the information on the small dam plans to the ground. This information provide lines, grades, and elevations for construction of the works in accordance with the plans. Consideration should be given to the contractor's requests to ensure optimum efficiency of the set out. The quality and appearance of the completed work will reflect the care and thoroughness exercised in the set out procedure (SCA, 1983).

The areas to be cleared usually will consist of the embankment site, the spillway site, the borrow pit, and the area over which water is to be stored. Each of these areas should be clearly marked with pegs. In the case of the storage area, the proposed waterline should be located accurately with a level. Clearing pegs should be at least 4 metres outside this waterline to give an indication of the area to be cleared around the edge of the storage.

The embankment is located by setting pegs along its centre-line at intervals of 10 metres or less. Usually this will have been done during the course of the initial planning survey. Fill and slope pegs are then set both upstream and downstream from the centre-line pegs marking the points of intersection of the side slopes with the ground surface.

The earth spillway is located by pegging the centre-line and then setting cut and slope pegs along the lines of intersection of the spillway side slopes with the natural ground surface. The procedure for setting these pegs is the same as for pegging the embankment, except that they are cut pegs rather than fill pegs. They should be offset so they will remain in place for referral during construction.

Where suitable fill material must be obtained from a borrow pit, it is essential that this area be clearly demarcated. Cut pegs should be set to control excavation within the limits of suitable material and to drain the borrow pit.

A spillway or trickle pipe should be located by pegs offset from the centre-line of the pipe and placed at intervals not exceeding 10 metres. The spillway should be located where it will rest on a firm foundation. Cuts from the tops of the stakes to the grade elevation of the tube should be plainly marked on the pegs. The locations of the low level pipe and gate valve, cut-off collars, outlet structure, and other appurtenances should be identified by clearly marked, additional pegs.

Setting out of a dam site can vary from the pegging of the Full Supply Level (storage area) at each end of the embankment to placing pegs all along the toes of the embankment and spillway and the placing of offsets. For small dams offset centre-line pegs (steel posts), full supply pegs (coloured) and one upstream toe peg are normally sufficient.

Centre-line pegs should be placed well away from the construction area. Two pegs placed 6 metres apart at one end of the bank enable the centre-line to be easily established during construction. Batter peg distances can then be calculated simply from a longitudinal centre-line survey. It is important that all pegs and the bench mark (see Section 4.2.2) are shown on the dam to minimise confusion.

The distance of toe pegs from the dam centre-line is calculated from a longitudinal centre-line survey as follows:

$$\text{Distance of toe peg} = (\text{Height at centre-line} \times \text{batter slope})$$
$$+ (\text{half of the crest width}).$$
$$= (6 \times 3) + \frac{3}{2}$$
$$= 19.5\,\text{m}$$

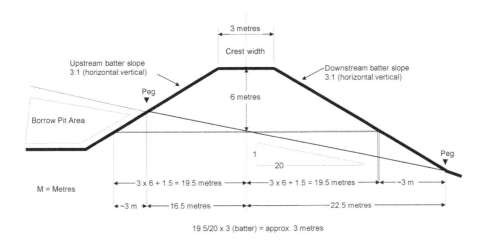

Figure 5.1 Location of toe pegs on slopes (*Source*: SCA, 1983).

This formula is only applicable to non-sloping sites. Adjustments must be made to correct for the slope, when measuring a site for a high dam or where steep slopes occur (for an example with 5 per cent slope [1:20, vertical : horizontal], see Figure 5.1).

In this example the correction is distance $= \dfrac{19.5}{20} \times 3 = 3\,\text{m}$ (approx)

5.4.2 Diversion of water

Provision must sometimes be made to divert water flowing through the site. If a pipe is being placed through the embankment, its early installation may be required to divert small flows during construction.

In wet drainage lines it is commonly necessary to construct a small coffer-dam (protective wall) to keep the site dry while the core trench is constructed. In this case, flows are either pumped or piped across the site.

5.4.3 Clearing and grubbing

Where the site is timbered, clearing and grubbing should be carried out to at least 6 m from the downstream and upstream toes of the embankment. Trees flooded by the dam should be cut off at stump height and then removed from the site. All holes made by grubbing under the embankment should be filled with sound material which is well compacted.

5.4.4 Stripping topsoil

All plants and silty soil should be stripped from the embankment site and 'borrow' area (the site from which additional soil is to be taken). The topsoil should be kept separate from subsoil, and stockpiled for use during the later stages of construction, such as covering the embankment and spillway.

5.4.5 Core trench

The core trench (cut-off trench) must extend at least 0.6 m into impervious material or 0.3 m into rock and extend for the full length of the bank. All water, mud, loose soil and rock must be removed before backfilling commences. The backfill should be specially selected and placed in layers of no more than 100 mm thick. If a central core is used, the core trench must be continuous with the core.

The core trench should be a minimum of 2 m wide, extend on both banks to above top water level, and reach down to impervious material below. This last factor may not be practical but it is essential in deep soils that crack during seasonal changes, as it will create a 'cut-off' zone to stop the cracking in the embankment.

5.4.6 Borrow pit material

Where the borrow material or soil has adequate clay content, an impermeable embankment can be constructed without having to 'zone' the embankment (see Sections 2.1.1, 3.1.1 and Figure 3.2). If the borrow material is highly variable, it is advisable to place the material with higher clay content (free from stones and decomposed rock) in a zone along the axis. Soil with lower clay content is then placed on either side, upstream and downstream, of this zone. It should be noted that silty and sandy materials are not suitable for construction of a watertight embankment.

The waterway (river, stream or creek) bed soils should not be used either as a borrow pit within a water storage or to increase the volume of the storage. To do so can create a low point that prevents environmental flows passing through a low level pipe within the embankment (see Section 3). This can also lead to seepage problems if the borrow pit is in close proximity to the dam wall.

5.4.7 Selection and placing of material

If the material is uniform, no selection needs to be made. However, in a great many cases the soil is variable and a selection of materials is found to be necessary. Where good quality, impervious material is in short supply, it should be kept separate and placed as a central core. In general, where the soil materials are variable, they should be placed as shown in Figure 3.2. It is sometimes necessary to remove very poor quality material. This can be placed in the downstream side of the bank by flattening the downstream batter.

Soil should always be placed longitudinally in an embankment to avoid lines of weakness across the bank. Bulldozer-constructed banks require material to be spread as well as pushed up.

The thickness of soil layers should not exceed the capacity of the compacting plant being used, because each successive layer must be bonded onto the previous layer. Where material has dried out it must be scarified and watered before the next layer is emplaced.

5.4.8 Spillway and outlet structures

A spillway should be designed and installed so that it is of sufficient capacity for a 1 in 50 year or 1-in-100-year storm flow, and at a level which maintains adequate freeboard. A spillway return should be constructed clear of the embankment toe and with the least possible disturbance to the natural ground in order to minimise erosion. To protect the spillway from erosion during continuous flows, a trickle pipe with inlet pit should be installed through the embankment.

A low level pipe (compensation pipe) of an appropriate size, determined by catchment area and rainfall characteristics, should be installed at the base of the embankment with cut-off collars at intervals along the pipe as indicated in Section 3.4.3. It is preferable to place the pipe in natural ground rather than in fill material. All material around the pipe must be carefully compacted, preferably with a mechanical rammer.

5.4.9 Batters and topsoil

After the embankment and spillway have been completed, all loose uncompacted material must be removed before topsoiling. Topsoil is spread onto all the batters and crest of the embankment and all the cut surfaces of the spillway. It is also sometimes necessary to cover the cut batters of the borrow pit if erosion is likely.

The thickness of topsoil necessary for erosion protection, insulation and promotion of vegetation cover varies with slope, environment and soil type. A common fault is an excessive thickness of topsoil on a steep batter that slides off with the first heavy rainfall event. A thickness of 50 mm is generally sufficient to establish a vegetative cover while a 200–300 mm thickness is required to provide an effective insulation of the embankment against excessive moisture changes (that is, shrinkage and cracks).

The upstream batter is compacted more than the downstream batter to give greater stability where the soil is wetter. Sandy clays are usually more stable under moist conditions than soils with a higher clay content.

A 'freeboard' of about 1 m is required to protect the embankment from over-topping. To minimise drying and cracking of the impervious embankment, it is recommended that most of the freeboard zone is constructed from silty or sandy topsoil. Increasing the minimum crest level required for freeboard by 5–10 per cent allows for settlement of the embankment crest. If a wide crest is required for vehicle access, make sure that wheel tracks do not form as water will pool in them and increase the risk of erosion.

Topsoil stripped from the borrow area and embankment foundation should only be used in the outer metres of the embankment, and never in lower regions. The higher proportion of silt and organic material protects against rapid moisture loss from the embankment, stopping it from cracking and providing an environment for rapid growth of ground cover. Seeding the embankment surface to protect against erosion is also recommended. All disturbed areas must be protected against erosion by planting and maintaining a suitable holding grass for the geographic area involved. The use of fertiliser and watering may be necessary particularly during the establishment stages.

5.5 COMPACTION

Compaction is the most important factor in achieving a stable, durable and solid earth embankment, which is resistant to the constant seepage of water through the soil. Notably, many dams fail because of poor compaction.

Compaction occurs when pressures are applied to the soil so that the individual soil grains are pushed together as air is expelled. Compaction in the field is directed at reducing the percentage voids to less than 5 per cent. When soils are saturated, the application of sustained pressure will also expel water from the pore spaces separating the soil grains, resulting in a measurable reduction in soil volume. This latter process is called consolidation.

When soil is disturbed, transported and used in construction without being compacted by machinery, it is loose and friable with a high proportion of air spaces between the soil grains. The soil mass will have high permeability and low strength. Over time, this material will settle and become denser. However, settlement will be uneven and deformation in a number of different directions may occur. This will result in extensive cracking of the surface and eventual failure of the structure. To protect against this, artificial compaction is used in the construction of roads, earth dams and embankments (including levees and graded banks) so that soil physical and engineering properties are modified. After compaction, the soil mass should be characterised by high soil strength and low permeability (the rate at which water will infiltrate and flow through the soil). In addition, vulnerability to settlement in response to repeated loading should be significantly reduced.

Compaction produces a uniform product and requires not only the appropriate equipment and specifications, but also an understanding of the inherent properties of the soil to be used. Therefore, before compaction is undertaken, the permeability, shear strength and density of the undisturbed soil should be known. In addition, because over-compaction can cause failure of the structure, soils need to be tested to determine the maximum possible compaction.

5.5.1   Compaction when constructing a dam

Good compaction relies on a range of factors including soil characteristics, soil moisture levels, the type of machinery, weather conditions and construction methods. The soil to be compacted must be moist but not too wet, and must be layered along the full length of the embankment at depths appropriate to the equipment used. Farm machinery and hand methods are usually only sufficient to compact layers less than 50–75 mm deep. Heavier equipment such as rollers can work with layers up to 200 mm thick, and should be used where large quantities and wide areas of soil are involved. In addition to compactive effort (roller pressure and number of passes), the moisture content in the soil is vitally important.

Each soil layer should be bonded to the previous layer by light scarifying along the axis. Rolling should be continued until the roller feet do not completely penetrate the soil. Large stones and lumps of partly decomposed rock, which cannot be broken down by rolling, should not be incorporated.

This information has been written for those who are constructing their own dam. However, it is recommended that expert advice is sought before taking any action.

5.5.2   Recommendations for compaction

Compaction must form a part of the construction of all embankments.

* If a bulldozer is the only plant used for compaction, the correct moisture content is critical. This is determined by a qualified engineer or geo-technical expert.

- The moisture content for bulldozer compaction is higher than that required for compaction by a roller.
- Standard Optimum Moisture Content (SOMC) is roughly equivalent to moisture content at the plastic limit in clayey soils.
- When using a scraper, it is generally impossible to work with material that is too wet. In these conditions, the equipment will bog when the moisture content is greater than 4 per cent over optimum.
- Embankments of 3 m in height or exceeding $3 \, m^3$ in volume should be rolled with a 'sheepsfoot' or 'pad' roller.
- Flat-wheeled rollers are not recommended for the construction of small dams.
- If a dry crust forms during construction it must be broken up, watered and mixed or removed from the embankment.
- If the soil is dispersive, soil moisture content during construction must be at least OMC and a sheepsfoot roller should be used.

5.6 SOIL MOISTURE

The dam embankment should be constructed from soil that is sufficiently moist to be pliable without crumbling, but not so wet as to excessively stick to, or flow away from machinery. A simple test is to roll a small ball of the soil into a 'rod' between the hand and a smooth, hard surface. If you can roll it into the thickness of a pencil (about 7 mm) then the moisture content is near optimum. Soil that is too dry will crumble. Soil that is too wet will stick to your palm. A soil of the correct moisture is critical for construction purposes. It will layer and compact more easily, reducing the tendency of the embankment to 'settle' after the dam has been built, and will hold water in the dam for longer periods.

If the soil is too dry it will contain a high percentage of air spaces, even after compaction. Dry soils will take up moisture easily to the point of saturation, losing the necessary characteristics of strength and impermeability. If the soil is too wet, it becomes soft and unable to be compacted effectively.

5.6.1 Adjusting soil moisture

In some situations, such as where dam walls are higher than 5 m or when the soil has very high or low clay content, it may be necessary to measure accurately, and adjust the soil moisture content during construction. Soil should be tested in a laboratory to determine its clay content. The required moisture content for dam building varies from about 10 per cent for soils with low clay content to 34 per cent for the heaviest clays.

If soil is spread in thin layers when conditions are hot and dry, the moisture content could be too low because of losses through evaporation. This can be overcome by minimising the length of time the soil is exposed, either by waiting for cloudy conditions, or by adding water to the soil during construction. The addition of water can be undertaken by:

- mixing water into the soil by cultivation with a disc plough or rotary hoe;
- irrigating as the soil is spread out on the embankment; or

- deep ripping and irrigating the soil before excavation, also known as 'borrow pit irrigation'.

Borrow pit irrigation is more economic than adding water directly to the construction surface. It results in more even distribution of water, and saves time by avoiding the necessity to water the construction surface between each layer.

Borrow pit irrigation is undertaken by initially ripping and ploughing the 'borrow' area before irrigating it thoroughly, and leaving the water to soak in over one or more days before excavation. If you are unable to add water artificially, you may have to wait until sufficient rain falls. However, the amount of rain needed varies widely. For this reason, you will need to dig one or more holes at least 2 m deep, before excavation starts, to help assess the soil moisture content. If you think rain may interfere with construction it may be feasible to install drains to direct water away from the site or to pump water from a hollow in the excavation area.

Since additional time and money are involved in manipulating soil moisture content under dry climatic conditions, it is recommended that you avoid dam construction during very hot summer periods.

5.7   ALLOWANCE FOR SETTLEMENT

Settlement in small dams is due to consolidation or saturation. Consolidation is the process of squeezing out the pore water by the weight of the embankment itself. Consolidation settlement can be settlement of the embankment or the foundations. In general, most consolidation settlement takes place during construction, particularly with coarse grained (silty and sandy) materials. Consolidated settlement will continue for an appreciable time after construction when materials, such as highly plastic clays, are used. This is particularly evident when the materials are wet during construction. Under these conditions, consolidation settlement of up to 5 per cent can be expected (SCA, 1983).

Compacted soils may exhibit a sudden settlement when saturation occurs in response to seepage flow. This may lead to settlement of the soil below the seepage line. The joint between the settlement soil and the overlying unsettled soil is a common location of tunnel failure in dispersive soils. There is no simple method of predicting the amount of saturation settlement. However, it is greatest in clay soils and least in sandy soils.

Settlement includes consolidation of both the fill materials and the foundation materials due to the weight of the dam and the increased moisture caused by the storage of water.

Settlement by consolidation depends on the properties of the materials in the embankment and foundation and on the method and speed of construction. The design height of earth dams should be increased by an amount equal to the estimated settlement. According to different construction techniques, this increase should not be less than:

Rolled fill (Sheepsfoot roller)	5%
Scraper placed	8%
Dozer placed	10%

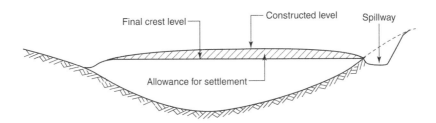

Figure 5.2 Settlement allowance.

5.8 EQUIPMENT

The type of equipment you should use for building your dam will depend on:

- the size of the proposed dam;
- soil type;
- the amount of soil and distance it is to be moved; and
- operating conditions.

i Bulldozer

Topsoil should first be removed from the borrow pit and embankment areas and stockpiled in the immediate area. Trees should not be included in the topsoil stockpile area and room needs to be made for the bulldozer to push the topsoil back over the embankment (that is, clear of any fences and trees). The topsoil stockpile should be graded so that any rainwater will not form ponds and make it too wet to push the soil material back onto the built wall.

Where there is a small amount of topsoil over uniform clay, it is better to construct the dam as shown in Figure 5.3. This technique allows the material to be pushed into place quickly because of the flatter grade. Suspect material should be dug out so that there is a layer of good clay connected to a sound foundation. Optimal compaction is achieved by running over each layer with a bulldozer.

The embankment should be built in progressive layers if a bulldozer is used. The excavation area usually needs to be ripped prior to pushing, with cross-ripping of the excavation area being necessary in hard conditions to gain greater depth and to obtain better broken material for pushing. As material closest to the proposed wall is pushed up to create the wall, dirt will spill around the sides of the blade and cause windrows. Successive amounts of material are then pushed between the windrows until there is a blade-full of soil available. This is then pushed up the wall and dropped when the topsoil is reached. Each ripped area is fully excavated.

The ends of the wall and corners are built first so that it is not necessary to push material a long way to complete the section. This also leaves the centre open to allow rainwater to drain away. It is important that all the poorer material is not utilised in one section of the main wall.

Steep gully edges, which would be under the wall, should be excavated to a 3:1 batter (horizontal:vertical). This will allow better compaction and prevent sheer cracks

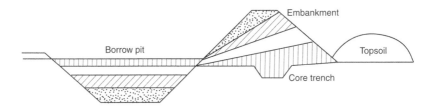

Figure 5.3 In situ layers of different soil types used to form an embankment.

forming if the wall material settles. A core trench the width of the tractor blade is then dug down to clay, and the trench is ripped and refilled with good clay.

Figure 5.3 is an example of how a bulldozer could be used to move individual layers of different soil types to form an embankment.

ii Scraper

Heavy earthmoving machines such as elevating or push-loading scrapers are not generally necessary for dam building unless you are in a hurry, short-haul distances are involved, or you can obtain the machinery at economic rates. A wheeled tractor or crawler-drawn dam scoop are sufficient for building some small dams. However, a vibrating roller is still considered the preferred machine.

5.8.1 Rollers

Rollers are designed for the purpose of soil compaction. They may have pronged feet or smooth drums, they may be vibrating or static, and be self-propelled or towed by a bulldozer. The type and size of the roller required depends on factors including soil type, size of the project and required thickness of the soil layers. The maximum recommended thickness of loose soil for compaction by a roller is 200 mm.

Although sheepsfoot rollers are the main technology in earthwork compaction they will not ensure success unless the roller is able to apply a sufficiently large pressure to destroy the clod structure of the soil and to produce a uniform distribution of density throughout each layer. If the foot pressure exerted by the roller is low, the lower sections of each layer will not be properly compacted and the roller will walk out after very few passes. Walking out does not necessarily mean the soil is compacted but only that the upper section is strong enough to support the roller. The lower section of each layer will in this case maintain their clod structure and possess large air voids which render the soil very permeable.

Vibrating rollers have much greater ground contact area and hence lower pressures. With the vibrator operating at a correct speed the contact pressure is increased by the dynamic load of the vibrating mechanism. This increase in pressure is greater on hard surfaces than on soft one.

Self powered compactors are sheepsfoot wheel types which is frequently a satisfactory substitute for a standard sheepsfoot roller. They have the advantage over other vibrating rollers that any vibrating effect is generated by the propulsion of the roller. Some adjustment of standard moisture content and layer thickness may still be required.

Table 5.1 Minimum number of sheepsfoot roller passes.

Dam water height (m)	Soil classification	
	GC, GM, SC, SM,CL	ML,CH, MH
3 to 6	4	6
7 to 9	6	8

Note:
- USDA Soil Classification (see Section 2.5.2 and Figure 3.2);
- dispersible material should be compacted with two additional passes;
- homogenous banks of dispersible material, such as ML, MH, and CH should not be greater than 6 m in height.

During construction (assuming the material is at optimum moisture content), the compaction for banks up to 3 m may be done by bulldozer or dozer-scraper, provided the layers do not exceed 100 mm. Where embankments are between 3 and 9 m, there should be at least four passes of a sheepsfoot roller over a 150 mm layer. A guide to the minimum number of passes is provided in Table 5.1.

5.8.2 Features of roller compaction

The most important features of roller compaction are shown below.

- Subsequent passes of the same roller have less compactive effect at the same moisture content. This is due to the increasing resistance to compaction with increased density.
- The optimum moisture content for the first roller pass may be slightly higher than for the second pass, and thereafter for each successive pass. Consequently, some drying can be advantageous.
- The speed of the roller has little effect on the compaction achieved.
- Different rollers are better suited to different materials. Vibrating rollers are best for sandy soils at low moisture contents; sheepsfoot rollers for loam and dry soils at low moisture contents; and pneumatic type rollers for clays at high moisture contents.
- No roller will compacts the soil adequately if the layer is greater than 400 mm thick in the loose state.
- The use of the heaviest roller that does not cause subgrade failure is always recommended.
- A sheepsfoot roller requires more passes.
- Compared with a sheepsfoot roller, a higher moisture content is required for compaction when using a bulldozer, because of the low loading intensity.
- Flat wheeled rollers tend to cause layering.

5.8.3 Other machinery

It is not generally recommended to use bulldozers to achieve soil compaction. Bulldozers are designed to excavate soils and, unlike rollers, exert a low pressure and

compactive effort on the ground. However, bulldozers can achieve sufficient compaction by thoroughly track-rolling the soil in layers less than about 50 mm thick,
depending on factors such as soil type, moisture content and weather conditions.
Four to eight passes are needed over the same area, limiting the amount of soil that
can be efficiently compacted by bulldozer alone. Nevertheless, very small dams made
from impermeable soil up to 2 m high can be constructed successfully by using only a
bulldozer.

Machines such as rubber-tyred scrapers or tractors exert greater pressure on the
ground than bulldozers but can be less effective than rollers. Moreover, soil that is
sufficiently moist for compaction by a rubber-tyred machine can cause the tyres to
slip. The distance between the tyres also means that extra passes of the machine are
needed after the soil has been unloaded. These machines should compact the soil in
layers no thicker than 100 mm. Tractor-drawn scoops tend to be slow and require a
tractor of at least 40 kw capacity. If using this type of machinery, remember to calculate
the length of time it will take to excavate and compact the soil, and allow for this when
scheduling the project.

5.9 INSTALLATION OF OUTLET PIPE

Seepage and piping along outlet pipes (see Section 3.4) are a major cause of failure of small dams. As pipes become larger in diameter it becomes more difficult to
compact material properly, down the sides of, or under, the pipe. Therefore, backfill
should comprise material of a similar moisture content to the in situ material to avoid
shrinkage cracks at the excavated interface. Mixing with expanding clay is also good
practice. Large pipes (450 mm diameter and above) require layer thickness no greater
than 100 mm (loose) to ensure adequate compaction by pneumatic or mechanical hand-
held rammers. Compaction and filling should continue until there is sufficient cover
to avoid disturbance and/or crushing of the pipe.

5.9.1 Testing of the pipe

After the pipe is in place, and before any backfilling, the pipe should be tested for
leakage, applying pressure through the use of compressed air. Testing should be carried
out at a pressure of 150 per cent of the maximum working pressure. No leakage is
permitted over a two-hour test period.

Where pipes that are prone to damage are used, such as rubber ring-jointed PVC
pipes, a further test is recommended after 1 m of soil cover has been emplaced. Failure
at this time can be corrected at a much lower cost than after completion of the
embankment.

5.9.2 Foundations on rock

Where rock or 'hard ground' is exposed in the embankment foundation, all loose,
shattered or flaky rock and soil should be removed. Any 'overhanging' rock sections
should be broken down to a 1:1 (horizontal:vertical) slope by barring or wedging.
Before any fill is placed in the core trench or in the embankment all exposed rock

should be cleaned and any holes left by loose rock should be filled with good quality CL material.

5.10 CHECKING FOR COMPLIANCE WITH STANDARDS

5.10.1 Checking the contractor's work

It takes time to set out a project properly at the start. Ideally, the designing engineer undertakes the dam layout, but if the contractor does it, you must expect to pay for it. The contractor should certainly check the work by taking levels during and after the project and this, too, will add to the cost. However, for you, it is money well spent (McMullan, 1995).

While the owner, engineer and the contractor have responsibilities for compliance with standards, inspection is also required if high standards are to be maintained.

5.10.2 Inspection during construction

There are many items in the construction of a small dam that need to be checked during construction. Responsibility must be assumed by the owner, since the designer of the dam may not always be able to spend sufficient time on the project. The landowner should be encouraged to watch the construction so that he/she can inform the contractor/engineer if corrective action is necessary. The following are items that might be checked by the landowner:

1 All clearing and grubbing operations should be completed according to specifications before any work on the embankment is started.
2 Before embankment construction begins, the foundation should be properly prepared. The completed cut-off trench should be inspected to ensure that it is excavated to impervious material and is free of water before it is backfilled.
3 The completed installation of the spillway, pipeline cut-off collars and other outlet structures should be inspected before embankment construction is started. Materials used, location, alignment, grades and dimensions should be checked for compliance with the plans.
4 The earth excavation and the selecting, placing, spreading and compacting of the material in the embankment should be inspected frequently to ensure that specifications are met.
5 The landowner may wish to photograph key stages in the project for record purposes.

5.10.3 Good work takes time

There are no short cuts available if you want a top quality project completed.

First, the dam site must be ripped. If the project is done well, you won't finish up with any hard patches on the soil surface, and it is important to note that quality of finish is evident in the hard to get at places. Good contractors tidy up the crest and batters, and especially the awkward corners (SRW, 1995).

The vital test of a top-quality project is the compaction of soil. It doesn't matter how good the equipment looks, or how big the tractor is, time must be spent achieving good compaction.

Defective dam-building includes the following practices:

i The use of inexperienced contractors and/or inadequate supervision

Nothing can take the place of a reputable contractor, experienced machine operators and appropriate equipment. For large dams or those that have specific problems, working from plans, specifications and material test requirements prepared by an experienced engineer is also vital. Even the best contractor may be tempted to take the occasional short cut in the absence of supervision.

ii Failure to strip topsoil from foundations, or otherwise incorporate in the bank

A large number of dams are still being constructed where the topsoil is incorporated into or left under the clay material that forms the bank. Such dams are more likely to leak, or ultimately fail, because topsoil is more permeable than clay and any organic matter in topsoil will rot allowing seepage to occur.

iii The use of inappropriate or defective materials

Small dams are built from different soil types ranging from sand to black clays. Each soil type has its own characteristics and problems that must be considered when designing a dam. The two soils most susceptible to problems are dispersive clays and pervious soils. A dispersive clay is one that erodes spontaneously in the presence of water. This is evident where local dams and creeks have a constant 'muddy' look. Swelling and cracking clays also present potential problems. Using these soils can result in dam leakage, which in turn can lead to piping or tunnel erosion. If unchecked, this will lead to rapid dam failure.

iv Inadequate compaction

This can result from using soils that are too wet or too dry during dam construction, or using tracked plant rather than rollers. Even one layer of inadequately compacted material in a bank can result in seepage and eventual dam failure.

v The incorporation of outlet pipes or other structures in the dam wall

Unless backfilling and compaction are carried out with extreme care, and cut-off collars provided, outlet pipes can provide an easy path for leakage. As far as possible, outlet pipes should be placed in the original ground rather than the embankment. Care is needed if such items are placed in the embankments or foundations of dispersive clays, as dam failure is highly likely unless the job is overseen by an expert.

5.10.4 Extra 'bonuses'

The conscientious contractor will think of 'extras' such as leaving a load of soil where stock troughs and channel crossings are to go. It helps if this soil is dumped at the right spot while the equipment is in the paddock. A good contractor will also think about where to put the topsoil and clay. Clay should be buried underneath topsoil, not the other way around! Some contractors will make it easier to spread seed on batters, and protect the batters against erosion, by 'roughening' the topsoil.

5.10.5 Changing your mind

Problems arise if you get 'bright ideas' halfway through a project. If you pushed the contractor hard on the original price, don't expect to get extra work done without an increase in the total cost. For example, if you have a plan of the proposed dam, it will show the contractor what to do. You create difficulties if you decide, when the job is half-done, that you want a vehicular crossing or trickle pipe constructed. Avoid changing your mind after you have agreed to the plan.

5.10.6 Progress payments

The contractor does not want a big overdraft any more than you do. When you are dealing with thousands of dollars, it is reasonable for the contractor to receive progress payments, similar to house-builders. It is your responsibility to negotiate agreed progress payments with your contractor (see Sections 1.9 and 3.6).

5.11 FINAL INSPECTION AND MEASUREMENTS

The final inspection by the design engineer should include sufficient profile and cross-section readings to ensure that the height, crest width, batter slopes and other dimensions shown on the plans have been met. Elevations of the top of the trickle pipe or spillway in relation to the control section of the spillway, cross-sections and profile of the spillway should be surveyed to ensure that construction has been undertaken in accordance with the plan dimensions and left in the specified condition. The final inspection should be made immediately after completion of the work and before the contractor moves his equipment from the site. All observations and measurements made during the final inspection of the dam construction should be recorded. Some State authorities responsible for licensing dams require a formal completion report and 'as constructed plans' on completion of works.

Maintenance

There is always the possibility that the dam you build might experience partial or total failure causing extensive damage downstream. This could result in loss of life, injury to people or livestock, damage to residences, industrial buildings or railways, and interruption to the service of public utilities (for example, electricity). In addition, loss of income resulting from lack of water could make the economic consequences of a dam failure substantial.

With newer small dams, surveys have shown that many fail due to lack of maintenance in the first few years after construction. Although owners enjoy the benefits of having a small dam, they also have to take total responsibility for the dam's safety and must, therefore, undertake adequate monitoring and maintenance.

This section sets out to explain some of the structural risks and hazards of the small dam. It also describes how to carry out routine monitoring and maintenance of a dam's condition. It is intended to cover all farm storages. The guidelines, if followed, could avoid costly repairs and extend the useful life of dams as well as decrease risks to the public (DC&NR, 1992).

6.1 SAFETY SURVEILLANCE

Safety surveillance of a dam is a program of regular visual inspection using simple equipment and techniques. It is the most economically efficient means of ensuring the long-term safety and survival of the dam. Its primary purpose is to monitor the condition and performance of the dam and its surroundings, and to ensure maintenance before the development of potential hazards.

The procedure for dam safety surveillance is unique to each dam but essentially it consists of:

- regular, close examination of the entire surface of the dam and its immediate surroundings;
- taking appropriate measurements; and
- keeping concise, accurate records of observations.

This inspection would normally occupy less than an hour. Frequency of inspection is discussed in Section 6.2.4.

For continuity and consistency of approach, the same person should normally carry out each inspection. An exception would be when an experienced engineer is brought in to inspect or advise on a particular problem.

6.1.1   Equipment for inspection

The following items are useful when conducting an inspection:

- dam inspection checklist to identify the items to be examined;
- notebook or diary and pencil – to write down observations at the time they are made, thus reducing mistakes and avoiding reliance on the memory;
- camera – to provide photographs of observed field conditions. Colour photographs taken from the same vantage points are valuable in comparing past and present conditions;
- shovel – useful for clearing drain outfalls and removing debris;
- stakes and tape – used to mark areas requiring future attention and to stake the limits of existing conditions such as wet areas, cracks and slumps for future comparisons;
- probe – a 10 mm diameter by 1 m long blunt-end, metal rod with right angle 'T' handle at one end. The probe can provide information on conditions below the surface such as depth and softness of a saturated area. Soft areas indicate poor compaction or saturated material. (An effective and inexpensive probe can be made by removing the head from a golf club); and
- hammer – to test soundness of concrete structures.

6.1.2   Observations to be recorded

All measurements and descriptive details that provide an accurate picture of the dam's current condition must be recorded. This information falls into three categories as shown below.

1   Location – any problematic area or condition must be accurately described to provide adequate evaluation. The location along the length of the dam should also be noted. For this purpose, a fixed reference peg at one end of the dam can be useful. The height above the toe, or distance down from the dam's crest, should also be measured and recorded. The same applies to conditions associated with the outlet or spillway.

2   Extent of problem area – record the length, width and either the depth or height, as appropriate, of any area where a suspected problem is found.

3   Descriptive detail – a brief, yet detailed, description of a condition or observation must be given, including:
   - volume of seepage from point and area sources;
   - colour or volume of sediment in water;
   - length, width, displacement, and depth of cracks;
   - soil moisture condition in the area, dry, moist, wet or saturated;
   - adequacy protective cover (topsoil);
   - surface drainage;
   - batter slopes steepness;

- presence of bulges or depressions on the slopes;
- where deterioration has occurred, describe whether it is rapid or slow; and
- changes which have occurred since the previous inspection.

This is not a complete list, but serves as a guide. If a condition has changed since the last inspection, records should be kept in the diary as follows:

- describe the change;
- take photographs of the site, including a close up of the changed condition, if possible; and
- note the date and a description of the scene shown in the photograph.

Remember, the primary purpose of the inspection is to pick up changes that have occurred since the previous inspection. If a situation looks as if it is worsening, or causing concern, the owner should not hesitate in seeking professional help. Table 6.1 is a typical checklist for noting defects and keeping long term records.

Table 6.1 Dam inspection checklist.

Areas of dam	Items to address	Observations
a. Upstream slope b. Crest c. Downstream slope	• Protection • Uniformity • Displacements, bulges, depressions • Cracking • Erosion • Rabbit, wombat or yabby activity • Obscuring growth (trees) • Wetness • Changes in condition • Stock damage	
Seepage	• Location • Extent of area • Characteristics of area (i.e. soft, boggy, firm) • Quantity and colour • Transported or deposited material • Spring activity • Boils • Piping and tunnel erosion • Changes in vegetation	
Outlet	• Outlet pipe condition • Operability • Leakage • Downstream erosion • Gate valve leakage	
Spillway	• Condition of downstream slope protection • Spillway obstructions • Erosion or back cutting in spillway	

Note: A sketch plan of the dam assists the recording of observations. Ideally it should be approximately to scale and the locations of observations should be indicated on it.
Source: DC&NR, 1992.

6.2 INSPECTION PROCEDURES

Some preparatory work before an inspection provides direction and purpose. This may include a review of the previous diary entries to note any areas which will require special attention. If the inspection aims to re-evaluate suspected defective conditions discovered during the last inspection, any available construction drawings should be examined first for a possible explanation of the situation.

The best results can be achieved by developing consistent records of observations. Following a specific sequence of observations during the inspection is a useful approach, such as:

- upstream slope
- crest
- downstream slope
- any seepage areas
- outlet, and
- spillway.

This will lessen the chance of an important condition being overlooked. It is also wise to report inspection results in the same sequence to ensure consistent records.

6.2.1 General techniques

The inspection is conducted by walking along and over the dam as many times as necessary to see every square metre, a person can usually obtain a detailed view for a distance of 3 to 10 m in each direction (Figure 6.1), from any given location. The distance will vary according to the smoothness of the surface or the type of material on the surface (for example, grass, concrete, rock, brush).

To cover extensive surfaces properly, several passes are required (Figure 6.2). Adequate coverage can also be achieved using parallel or zigzag paths (Figure 6.3).

On the downstream slope a zigzag path is recommended to ensure that any cracking is detected. At several points on the slope, you should stop and look around through

Figure 6.1 Sight distance.

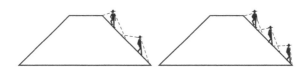

Figure 6.2 Successive passes.

360° to check alignments and ensure that some important feature of the slope has not been overlooked.

6.2.2 Specific techniques

Monitoring changes that will occur over a period of time can include a number of methods, including:

Sighting

A sighting technique, similar to that used when selecting straight pieces of timber, can be used in identifying misalignment as well as high or low areas along a surface. This technique is illustrated in Figure 6.4.

The same method can be used to sight along the crest of a dam. Centre the eyes along the line being viewed. Sighting along the line, move from side to side a little to view the line from several angles (Figure 6.5). Looking through a pair of binoculars will make any variations more obvious.

Probing

The blunt end of the probe (see Section 6.1.1) is pressed into the earth on the batter slopes, crest or at places being inspected. Conditions below the surface, such as depth and softness of a saturated area can then be noted. By observing the moisture brought up on the probe's surface and the resistance to penetration, it is possible to decide whether an area is saturated or simply moist.

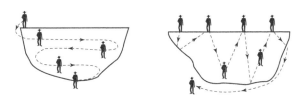

Figure 6.3 Coverage path.

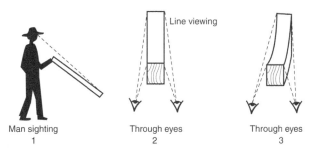

Figure 6.4 Sighting technique.

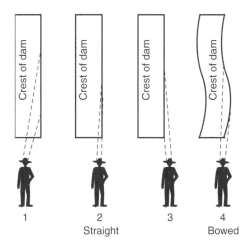

Figure 6.5 Sighting along crest.

Pegging-stakes

The best way to find out if there is a leak is to check how fast water is disappearing from a storage by marking the waterline with a peg at regular intervals, for example weekly. If the storage is used for stock or irrigation, the waterline should be pegged before and after use.

Measuring in this way is much better than simply guessing. A suspected leak, when measured, may turn out to be only evaporation loss. Evaporation can easily be 5 mm per day, and as much as 10 mm per day in dry and windy conditions.

Uneven surfaces and displacements

Slides are as difficult to spot as cracks. Their appearance is subtle, since there may be less than 1 m of depression or bulging in relation to the normal slope for a distance of perhaps 10 m. This may be complicated further in situations where the dam was not uniformly graded by the bulldozer operator during construction. Therefore, familiarity with how the slope looked at the end of construction helps identify any slides. One method of monitoring batter surface movement is to place a straight line of stakes down the slope with a string tape attached to the top of each stake. The point at which a slide takes place will cause the uphill stakes to be uprooted, whilst those just downhill of the movement will show a slackening of the string tape.

6.2.3   Evaluation of observations

The record of observations taken at periodic inspections is used to develop a mental picture of the dam's performance over time. Accurate measurements pay off here because small changes, which could go undetected if simply looked at, will show a pattern or trend.

Immediately following the inspection, the observations should be compared with previous records to see if there is any condition, measurement, or trend that may indicate a developing problem. The owner can then begin to address any potential problem before it becomes a threat to the dam. When a significant change is detected, any design drawings for the dam should be examined carefully to see if an obvious reason for the change is apparent. If there is any uncertainty regarding a change, a professional engineer experienced in the field of small dam engineering should be engaged immediately to determine if any action, such as increased monitoring or detailed investigation of the condition, is required. These actions will help ensure the safety, safe operation, and long, useful life of the dam.

6.2.4 Frequency of inspection

The required frequency of inspection required for an effective program of monitoring is dependent on a variety of factors including:

- size or capacity of the dam;
- condition of the dam (risk category); and
- potential for damage resulting from failure of the dam (hazard category).

The larger a dam, the more frequently inspections are required. Notifiable dams that have a significant hazard category, for example, need to be inspected more frequently than once a year. Dams in this class are defined as being >5 m in height with a 50 ML storage capacity, or 10 m in height with a 20 ML storage capacity and their failure would result in significant economic damage or loss of life (see Section 10.4.1).

The inspection frequency for a particular dam is the responsibility of the owner, although large dams, or those suspected of being in a high risk or hazard category, require professional advice. Detailed, comprehensive inspections can be alternated with more frequent, rapid, visual inspections aimed at detecting unusual changes that have occurred in the interim period.

Following a regular routine like this should enable the dam owner to become aware of faults before partial or total failure occurs. Times when additional close inspections are recommended are:

- before a predicted major rainstorm (check spillway and outlet pipe);
- during and after severe rainstorms (check spillway and outlet pipe);
- during and after a severe windstorm (check upstream batter slope for damage from wave action); and
- after any earthquake or tremor, whether directly felt on the owner's property or reported by local news media.

Inspections should be made before and during construction, and also during and immediately after the first filling of the storage.

6.3 CAUSES OF DAM FAILURES

There is little that a dam owner can do to prevent dam failure in response to earthquakes, extreme storm activity and failure of upstream dams. However, normal margins of safety should be capable of accommodating earthquakes of a magnitude that is appropriate for the region, based on geological information. Advance emergency planning to minimise risk and uncertainty is the only measure available to the dam owner (Lewis, 1995a; DC&NR, 1992).

Ingles (1984) observed that failures fell into the following categories, as follows: overtopping 20%, piping 26%, slope failure 20%, foundation failures 17%, all other 17%. Overtopping occurs when the actual flow over a spillway exceeds the flow from which it has been designed (see Section 3). It may therefore be regarded as a 'natural hazard', resulting from extreme low probability weather conditions. The other main types of failure listed may be regarded as related to human error. The high percentage corresponds broadly with overseas observations. Of these human-error-related failures, two-thirds (piping and slope stability) are more related to improper construction and operation controls. One-third (foundation failures) are more related to errors of judgement in design and geological assessment (see Section 2).

Effective monitoring and maintenance programs can protect dams against the main causes of failure. A systematic program of safety monitoring should ensure detection, in good time, of processes that require major repairs or cause dam failure. Typical problems, which can be remedied if detected early enough, include:

- seepage/leakage
- cracking
- erosion, rilling and piping
- deformation and settlement
- concrete structure defects
- spillway blockage/erosion, and
- animal damage.

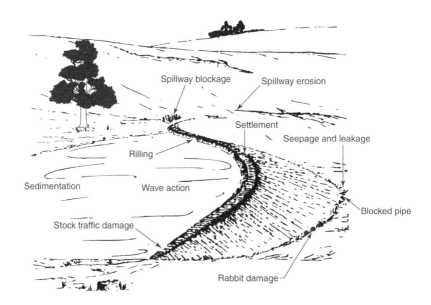

Figure 6.6 Problem areas on a dam site.

6.3.1 Dispersive clays

Dispersive clays are common in many parts of Australia. They occur in soils whose clay minerals separate into single grains when placed in contact with water. Typically, water is cloudy or turbid where dispersive clays occur. Dispersive clays are associated with high soil erodability and their distribution often coincides with the occurrence of erosion gullying, rilling and piping. Dispersive clays in a dam embankment can result in the leaching out of material from the embankment with consequent erosion and failure. The failure may be gradual but often occurs very rapidly, with little or no warning. A significant proportion of such failures has taken place during the first filling.

As far as possible the use of dispersive clays in dam construction should be avoided, but when this is not possible, the addition of a small proportion of lime or gypsum, well mixed in with the embankment material, can help to stabilise it. In existing dams, application of these chemicals to the surface layer of the upstream face of the embankment may be beneficial. As a guide, application rates of 3–5 per cent well mixed with the upper 200 mm would be used, but professional advice should be obtained.

Compaction of dispersive clays must be carried out to very high standards in accordance with specifications prepared by an experienced engineer. The moisture content during compaction must be carefully controlled to be on or marginally above, the optimum level. This can only be determined by laboratory testing of the material.

6.3.2 Seepage and leakage

Water escaping from the dam can occur locally or over a wide area. It might be visible on the embankment, the downstream toe of batter slope or the abutments, and be either clear or muddy. The rate of flow might be small or large, steady or increasing. Unless the flow is clear and the rate only small and not increasing, most forms of leakage represent a warning of potentially serious problems and indicate the need for early remedial work. The known or suspected presence of dispersive clays in the embankment or foundations would be an indication for even greater urgency. It is important that the embankment is well maintained and grass is kept relatively short so that seepage is readily identified if it occurs. Unless the cause of seepage and leakage is readily apparent and the repairs immediately effective, expert professional advice should be sought.

6.3.3 Cracking and movement cracks

During dry periods there will always be minor cracking as the embankment dries out. This may be prevented by good topsoil and grass cover. However, some soil types are more prone to cracking than others. Where these types of soil are common, cracking is often a serious problem.

Transverse cracks running across an embankment can promote seepage. Longitudinal cracks running along the embankment can fill with water during a storm and saturate lower layers. This may cause part of the embankment to slump.

Ideally, large cracks should be filled as soon as possible with compacted clay, preferably mixed with expansible clay (with trade names of Bentonite or Volclay). In practice, this can be difficult and it may be necessary to trench out the cracks before filling them so the clay can be compacted.

It is well known that when highly plastic clays are subjected to alternating wet and dry conditions, they will alternately expand and contract and develop thin shrinkage cracks on the surfaces of dams. These cracks rarely exceed 600 mm in depth, and consequently with properly designed freeboard allowances are unlikely to be a serious problem. However, some failures have occurred in cheaply built dams that are less than 3 m high.

Tensile cracking is considered to be the most dangerous form of cracking in terms of dam failure. Tensile cracks are caused by differential settlement strains induced within the bank by earth movements which are, in turn, products of non-uniform settlement or consolidation.

The extent of this earth movement or deformation depends mainly on the relative compressibility and geometry of the earth dam, the foundations and the abutments. The influencing factors can distort the dam in different directions and so create a variety of cracking patterns that may be continuous or local. These cracks can be up to 150 mm wide. The two most common forms of tensile cracking are longitudinal and transverse (Figure 6.7).

Longitudinal cracks develop roughly parallel to the centre-line of the dam and are always in a vertical plane, which distinguishes them from shrinkage cracks, which develop in a plane at right angles to the soil surface (compare Figure 6.7).

Earth movements creating longitudinal cracks result from:

- differential settlement in zoned dams
- foundation settlement
- foundation heave.

Longitudinal cracks are not usually dangerous but have proved troublesome when combined with other weaknesses.

By contrast, transverse cracks need careful monitoring and can be particularly serious because they tend to run straight through the dam. The main causes of this type of cracking are:

- differential settlement due to steep and/or incompressible abutments;
- settlement of foundation; and
- saturated settlement of embankment.

Where saturated settlement of the embankment occurs, this will be due to construction of the dam with material that is too dry. Consequently when the storage fills, the dam material becomes saturated and then slumps. However, the material above the seepage line remains dry and firm, resulting in an arched-type failure.

6.3.4 Erosion

Erosion is also a problem with many causes and forms; the presence of dispersive clays will usually increase its severity. The following are among the most common forms of erosion associated with small dams:-

i Rilling of the bank (small erosion gutters down the bank)

This usually happens where there is no topsoil on the bank, and run-off becomes channelled. To rectify the problem, pack rills with grass sods or cover the bank with

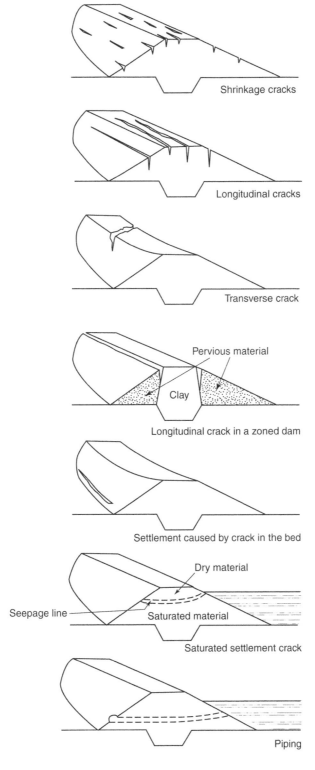

Figure 6.7 Different types of cracking in an embankment.

Photo 6.1   Longitudinal cracks on the crest of the embankment (*Source*: B. Lewis).

topsoil and sow down. Maintain a good grass cover on the embankment, and exclude stock.

ii Damage from wave action

Wind across the water creates wave action damage (see Section 3.1.5). Protecting the bank with rocks, hay mulch and netting, or grasses such as kikuyu can help. For larger dams, more substantial rock protection is warranted and the use of wind breaks should be considered.

iii Spillway erosion

By far the cheapest form of spillway is an earth type spillway (see Section 3.3.1), comprising either soil clay material or grassed, well-compacted earth. In the majority of cases, the nature of the site is such that an adequately sized spillway can be cut into the side of a hill without too much trouble. Where the site, type of soil, or extremely high flow prevents a normal 'earth spillway' from being constructed, concrete-lined spillways (either on an abutment or over the centre of the dam, see Section 3.3.4) can be provided.

A spillway return should be designed to spread the flow out to as great a width as possible and to keep water flow velocity to a minimum. The spillway return slope is that section of the spillway that returns the dam overflow to the creek. This section is also subject to erosion. Consequently, it should also be grassed and, preferably, should cross over natural, undisturbed ground. The gradient should be as flat as possible. The spillway return should be located to direct flow away from the batter toe of the dam to avoid erosion during floods.

Heavy flows over spillways can lead to erosion. Where this occurs, minor erosion should be filled with grass sods, covered with hay mulch and pinned down with plastic netting. The spillway should remain as level as possible across its entire width. The installation of a trickle flow pipe (see Section 3.4.1) with a pit will protect the vegetation cover on the spillway from prolonged saturation following a storm. This is not an alternative to the outlet pipe. If erosion persists, more substantial protection is probably required.

iv Sinkholes

Sinkholes are holes or depressions at the surface resulting from internal erosion that has caused underground cavities into which the surface material eventually subsides. Sinkholes are often a sign of severe and widespread hidden damage caused by processes such as piping. It is necessary to determine the nature and trace the extent of such damage, and to backfill all eroded areas with well compacted (non-dispersive) clay before dealing with the visible surface holes or depressions in a similar manner. Unless the defect is found to be superficial and associated with surface run-off rather than leakage from the dam, professional advice should be sought.

v Wind erosion

Erosion due to wind action can occur when the dam wall material has a high sand or silt content, and vegetation cover is poor. It is often associated with the passage of stock. Re-establishment of good grass cover is the best remedy.

A fence to exclude stock should be constructed around the perimeter of the embankment, storage area, spillway and spillway downstream slope as soon as possible after the dam is completed. Apart from damaging the grass cover and creating 'cattle pads' which can lead to serious erosion, the water in the storage can become turbid and polluted by continued stock access. The best alternative is to provide stock water at a trough remote from the dam, or install a fenced walkway to a restricted area of the dam.

vi Rabbit and wombat damage

Animal burrows can be a source of seepage. They need to be dug out and repacked with clay. If the burrows are extensive, the storage should be emptied before they are dug out, or professional advice sought. Effective rabbit control needs to be maintained.

vii Erosion at outlet valve

Erosion can be caused by water discharging through low level pipes beneath a dam. Provision should be made at the outlet pipe point of discharge to reduce the velocity of exiting water. It is recommended that rock with minimum diameter of 300 mm should be placed on a layer of crushed rock to minimise erosion, or an effective outlet structure constructed to dissipate the energy of the discharging flow.

viii Piping

When a failure is the result of an internal pipe or tunnel forming through an earth dam, it is usually referred to by contractors as 'piping' and this may be a result of a number of causes. The three most common causes are given below (Figure 6.7).

1 Conventional piping

Water seeps through all earth dams, whether large or small. It is not the occurrence of seepage that is problematic, but the rate of seepage. If the rate is high, water will have the capacity to mobilise and transport soil particles. Viscous drag forces within the dam will oppose this movement of water and so reduce the eroding forces operating through the soil. If the drag forces exceed the eroding force no piping will occur, but if they do not, then there is a serious likelihood that piping failure will occur. Any line of weakness within the earth dam can accelerate piping failure. This process may develop because no dam can be built of truly homogeneous (identical) material.

Conventional piping occurs in cohesionless soil (such as a silt). This form of piping starts at the downstream side of the dam and then slowly proceeds to the water face upstream. Collapse of the pipe also accelerates surface erosion.

2 Tunnelling

Tunnelling occurs in dispersive soils, which are soils that can be broken down or separated into single grain components when placed in contact with water. The process is initiated when water enters the dry soil through a crack, and then disperses the soil exposed on the sides of the crack. The dispersed particles are carried back, in suspension, to the dam storage. This process continues as more water comes into contact with fresh dry soil and it will gradually take the form of a tunnel (or pipe) which will eventually develop through the dam.

Since tunnelling starts at the water side of the dam it is more difficult to detect than conventional piping, which starts on the downstream face. This makes tunnelling far more insidious and dangerous than conventional piping.

3 Leaking pipe

Another form of piping results from poor laying techniques used on outlet pipes as described in Section 5.9. This includes poor compaction around, and particularly beneath the pipe, which is a difficult zone to reach and is frequently neglected. The disturbance of a laid pipe by the movement of earthmoving plant over it can also initiate problems. For example, a dozer may hit a cut-off collar (see Section 3.4.5) and this in turn may create a 'roofing' condition (Figure 6.8).

A frequent cause of outlet pipe failure is leakage in a poorly jointed pipe. This is usually accelerated where a downstream valve has been included in the dam design (Figure 6.9). The use of an upstream valve would at least decrease the pressure on the joint and ensure that this reduced leakage flow would be confined within the pipe.

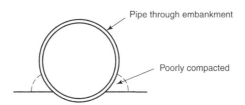

Figure 6.8 Pipe bedding problems.

6.3.5 Deformation and movement

i Settlement

Settlement occurs in response to poor compaction (see Section 5.7). Monitoring for settlement includes checking the distance between full water level and the top of the bank (the freeboard). Hard and fast rules cannot be laid down, but as an indication, it should not normally be less than about one-fifth the height of the embankment, or an absolute minimum of 0.5 m, whichever is the greater. If it is less than this, the embankment should be raised or the spillway lowered.

ii Slides

Slides can be described as the downward movement of soil on the slopes of dams. The slide usually has three distinct features. First, tension cracks develop on the crest, then a substantial portion of the dam slips downwards, and finally a pronounced heave occurs near the toe. Warnings of likely slides are first given by these surface cracks on the crest or on the upper sections of the sloping sides (see Figure 6.7).

Material that is likely to slide can occur on upstream and/or downstream batter slope sides of an embankment. Slope failure and slides occur where dams are built:

- on soft or clayey foundations with low strength;
- where batter slopes are too steep; or
- where the bank is insufficiently compacted.

This type of failure is most dangerous because it usually occurs when the dam is full and therefore can cause a catastrophic flood wave downstream of the dam. Additional problems can arise if a slide obstructs the outlet pipe, because it would be impossible to release stored water. The danger, if detected early enough, can sometimes be controlled by the reduction of water level in the storage, or by the placement of stabilising rock fill at the toe of the potential slide.

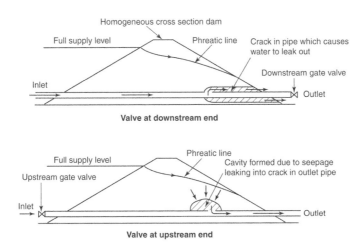

Figure 6.9 Pipe failure above and below gate valve.

Photo 6.2 Slump extends to the centre-line of the embankment.

Slides are major structural defects, normally requiring major remedial works such as flattening of batter slopes, improved drainage or the addition of rock-fill as a stabilising weight at the toe of a slope. Selection of an economical and effective remedy normally requires expert professional advice. In the short term, it may be necessary to drain or pump out the stored water.

Slides can be divided broadly into three categories.

1   Construction slides which develop during the construction phase and may occur on either the upstream or the downstream slopes. Almost without exception, construction slides occur where dams are being built on soft or brittle clayey foundations. This type of slide failure is seldom responsible for serious damage because it takes place before the storage is filled, but it can be embarrassing to both the

Photo 6.3 Slump on the downstream batter slope.

designer and contractor. Two forms of construction failure can occur. In the first case there is a rapid downward movement of soil over a very short period (from 5 to 15 minutes). The vertical drop of the slide could easily be as much as 3 m on a 6 m dam. In the second case, there is a slower, uniform movement spread over a much longer period, usually up to about two weeks, followed by a very slow creep. These two kinds of construction slides are due to differences in the soil structure of the foundation.

2   Downstream slides which may develop after construction. The two kinds of downstream slides are the deep slide and the shallow surface slide.

- Seepage through or under the dam creates internal pore water pressure causing a slide which moves deep into the clay foundation and often develops into the upstream slope of the dam. A frequent complication arises when it completely obstructs the outlet pipe making it impossible to release water as an immediate response.
- The shallow surface slide is often more due to saturation of the downstream slope after heavy rains. It may vary in thickness from 50 mm to 1.2 m and is most frequently found in poorly compacted small dams. It is not generally dangerous but can be expensive to repair.

3   Upstream slides which usually occur after a rapid draw-down of water in the storage. Like the deep downstream slide, it has a slide surface that cuts well into the foundation. Rapid draw-down refers to any water level drop in excess of

1000 mm per day (Table 3.1). Dam engineers make special allowances in their design for storages where rapid draw-down is anticipated.

Also hillside dams can be impacted upon by uphill slides, even though they are not located on the structure itself. In these cases, the balance between equilibrium and landslide conditions may be particularly sensitive on the slope, causing the least disturbance to mobilise vast volumes of soil downhill. The excavation for a hillside storage, together with the presence of groundwater, sometimes creates conditions resulting in an uphill slide into the storage.

6.3.6 Defects in associated structures

i Spillway blockage

The construction of an earth dam is not just a matter of pushing up a wall across a gully. Many dams built this way fail because they do not have an adequate spillway capacity to prevent floodwater overtopping the dam, or erosion on the spillway return cuts back into the downstream sloping batter. As with most problems, prevention is a cheaper and simpler strategy, so the following points are recommended when designing your dam.

- All too frequently the importance of an adequate spillway is not fully appreciated. When considering the size of a spillway, many people try to relate it to the normal everyday winter flows that occur in the creek or stream. What they do not realise is that these flows can be exceeded many times over and that the safety of the embankment, on which they spend so much money, depends entirely on the ability of the spillway to operate under high storm flow conditions.
- The costs of repairing a failed storage and paying compensation for downstream damages will be far greater than the cost and effort involved in providing a few extra metres of spillway width.
- If spillways are blocked in order to increase the storage capacity, the bank may overtop. This is a situation fraught with danger and must be rectified. Debris, bushes, trees, shrubs, fences and tall grass should be regularly cleaned from all parts of the spillway, including the approach area.

ii Outlet pipe blockage

The cleaning of an outlet pipe is a problem if the gate valve has been closed over the winter since accumulated trash, fish and siltation can clog the inlet on the upstream side of the embankment. This type of blockage can place a dam owner in conflict with the public regulatory authorities since approval for the construction of a dam on a waterway sometimes requires the gate valve (low level pipe) to be operated to ensure flows through the drier months of summer. These conditions cannot be met if the outlet pipe is blocked. Regular flushing is therefore recommended to minimise the risk of blockages.

In the event of a blockage there are different techniques and types of equipment available to dislodge an obstruction, including cleaning rods and flexible sewer pipe cleaners. However, most equipment can only be used when the storage is empty or close to it.

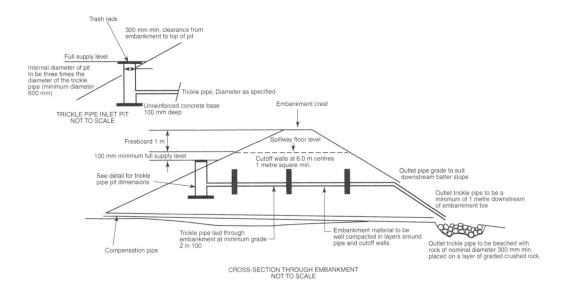

Trash rack

300 mm min. clearance from
embankment to top of pit

Full supply level

Internal diameter of pit
to be three times the
diameter of the trickle
pipe (minimum diameter
600 mm)

Trickle pipe. Diameter as specified

Unreinforced concrete base
100 mm deep

TRICKLE PIPE INLET PIT
NOT TO SCALE

Embankment crest

Freeboard 1 m

Spillway floor level

100 mm minimum full supply level

Cutoff walls at 6.0 m centres
1 metre square min.

See detail for trickle
pipe pit dimensions

Outlet pipe grade to suit
downstream batter slope

Outlet trickle pipe to be a
minimum of 1 metre downstream
of embankment toe

Trickle pipe laid through
embankment at minimum grade
2 in 100

Embankment material to be
well compacted in layers around
pipe and cutoff walls

Outlet trickle pipe to be beached with
rock of nominal diameter 300 mm min.
placed on a layer of graded crushed rock.

Compensation pipe

CROSS-SECTION THROUGH EMBANKMENT
NOT TO SCALE

Figure 6.10 Trickle pipe for a small dam.

iii Trickle pipe

A trickle pipe is a high-level outlet pipe that will discharge before the earth spillway comes into operation (Figure 6.10).

To ensure the best performance of a spillway, small flows on a continuous basis should never be allowed to pass for long periods (several days). These small flows are more likely to cause erosion than much larger flows of short duration. Dams supplied by springs are particularly subject to erosion of this type. To minimise this problem, a trickle pipe can be installed to carry the low flows through the dam without crossing the earth spillway. The size of the trickle pipe will depend on the volume of these low flows.

iv Trickle outlet construction

Where erosion of the spillway by perennial or ephemeral low flows is an issue, a trickle outlet can be simply constructed to minimise the risks. The information provided, however, should be used as a guideline only, since widely varying circumstances can apply on individual properties.

A simple and cheap drop inlet pit can be made from a standard 400 mm × 400 mm concrete sewage distribution pit. These commonly have two opposite outlets to fit 110 mm pipes (the outlet size may need to be adjusted). For larger small dams the trickle pipe could be a 150 mm diameter pipe with a 1 m diameter drop inlet pit. Alternatively a 200 Litres (44 gallons) drum can be used as a mould to make the drop inlet pit.

The trickle pipe is laid under the dam crest and run down the downstream slope of the dam in a trench. Cut-off collars are used to prevent water flowing along the trench outside the pipe.

The larger the trickle pipe, the larger the flow that can be passed and the greater the cost of installation. For pipes up to 225 mm in diameter, PVC sewerage piping is generally used because it is cheaper than steel, although concrete pipes are used in some cases. The cover over the pipes should be not less than 350 mm deep. In selecting a pipe class, it is advisable to contact the manufacturer to determine the most suitable pipe and depth of cover to support anticipated vehicle traffic loads.

The outlet pipe on the downstream side of the storage should direct water onto a rock mattress which is composed of graded layers of crushed rock to dissipate energy and prevent scouring in the creek.

6.3.7 Vegetation

Trees

Planting trees and shrubs can provide windbreaks that prevent wave action and asso-ciated soil erosion, and provide shelter for wildlife. Self-sown plants, trees or other deep-rooted plants should not be permitted on the embankment and spillway. The roots of these types of vegetation can provide a path for leakage through the dam, and ultimately result in its failure. Trees that are alive and apparently vigorous, such as poplars, can have dying roots with similar consequences. Vegetation larger than small shrubs should not normally be allowed to grow. Once established, a decision on whether trees should be removed, left undisturbed or severely pruned has to be made on the basis of the particular circumstances of the site. Rotting tree roots can also form tunnels that allow water to leak and seep through the dam bank. These tunnels can lead to failure through piping erosion. Seepage flow through the embankment or abutments can lead to the development of high water pressures near the downstream face, causing failure through sliding or erosion. Adequate compaction and the use of appropriate materials in the design and construction phases may prevent these condi-tions from occurring. If trees are required at a dam site, they should be planted at the upstream section of the water storage, well away from the dam and spillway, in line with the dam and at right angles to the dominant wind direction.

6.3.8 Total catchment protection

To maintain the required depth and capacity of a small dam, the inflow should be rea-sonably free from sediment. The best protection is to control erosion of the surrounding catchment area. Land with a permanent cover of vegetation, such as trees or peren-nial grass, provides the ideal catchment condition. If the catchment is not suitably vegetated, you may need to utilise cultivated areas that are protected by conserva-tion practices, such as contour tillage, strip-cropping, conservation cropping systems, vegetated desilting areas and other land improvement practices.

Desilting

While water levels are low in a dam, farmers have an excellent opportunity to remove excess silt and mud from storages. One simple method of desilting is to use a mud scoop of 1–2 m³ capacity and two tractors. The mud scoop is pulled through the dam by the cable of a heavy crawler tractor and the mud is deposited on the downhill side.

The empty scoop is returned to the other side by a lighter tractor and then pulled through as before (Nelson, 1968).

Care should be taken to ensure that the equipment does not expose sandy seams on the sides or bottom of the dam as these would cause excessive seepage. If sandy seams are discovered, leave 150 mm of silt to form a water-tight layer. High levels of silt in a dam indicate soil erosion problems within the catchment.

6.3.9 Weed control

Aquatic weeds in small dams can block pump and pipe inlets, deter stock from drinking, and in some cases, taint the water (see Section 8.2). If weeds are treated when they first appear, dams can be kept relatively free of some of the more troublesome species. All plants can become a problem and each type may require a different control method. However, in all situations the same factors should be considered in deciding what control methods, if any, should be used. In each case:

- determine whether there is a problem, and if so, what it is;
- identify the plant causing the problem;
- find out what control methods are available and which of them could be used safely;
- investigate whether these control measures could cause any other problems (for example, toxicity to fish and livestock) and if so, whether they can be avoided; and
- decide whether control is practical, desirable and worthwhile.

6.4 DAM LEAKAGE

There is nothing more frustrating than to find that the dam you built to water the stock and the house garden, to grow some fish, and to use for fire control becomes a muddy puddle when you most need it. But many dams do just that. Anecdotal evidence suggests that approximately 35 per cent of all newly built dams fail, primarily because of soil type and the construction technique. Dams are built in a range of soil types from sand to the stickiest of black clays. Each general soil type has its own characteristics and problems. In building construction you can designate the quality of the concrete, the timber and the steel, but with soil, you can't be so sure. In fact, no dam can be guaranteed against leakage or failure. However, the risk can be reduced to nearly zero by taking the right precautions.

Dams can leak simply because the construction was faulty. Where the bank is built on top of the natural surface (that is, it is not compatible with the sub soil), a line of weakness at the base of the bank allows seepage of water.

How a leaking dam is fixed depends on why it leaks. Therefore, the same solution will not work every time. Leaking dams are usually fixed by one of four methods:

1 rebuilding;
2 the use of soil additives;
3 the use of artificial liners; or
4 the use of clay liners.

6.4.1 Rebuilding

Badly built dams will always give problems and no amount of chemicals or liners will overcome inherent design or construction faults. Some temporary works may help in the short term in some instances but usually these dams eventually need rebuilding.

Since compaction is one of the most important factors in dam construction, it should be remembered that wheeled tractors or sheepsfoot rollers are superior to bulldozers. This is largely because bulldozers are designed to spread their weight over the tracks with a large surface area and therefore exert low pressure. With the cost of bulldozing being quite considerable these days, it is important to get any reconstruction job done by an operator with a good reputation for building dams (see Section 5.8). The best approach is to ask neighbours with good dams 'Who built your dam?' The local water distribution officer is also a good source of information. Two or three quotes should be obtained for the project, including details on when construction can be done. Remember contractors, like farmers, are often at the mercy of the weather and usually can not guarantee to be at your property on a particular day.

6.4.2 Use of soil additives

Soil additives are used to stabilise some dispersible clay soils and to clog up other soils that are highly stable but leak (SRW, 1995).

i Gypsum

Because of its calcium content, gypsum has been recommended to stabilise dispersible clay and hence reduce the incidence of tunnelling and slumping. Gypsum should be added to the soil when building new dams and after repairing failed dams in dispersible clay soil areas.

ii Bentonite or Volclay

Bentonite or Volclay are trade names for a naturally occurring clay that swells up as a gel, when wet, to about ten times its original size (CETC, 1995). It is this property that makes it useful for sealing dams. These expanding clays are best suited to reducing seepage in dams made from sandy or gravelly soils. Where it is impractical to drain the dam the clay can be sprinkled onto the water surface of the storage. Although it has a role in emergency repairs, it is only of limited use. If the leak is too rapid, or the clay is too thinly distributed in the water, the effect will be minimal.

iii Sodium tripolyphosphate (STPP)

This chemical has the opposite effect to gypsum. It aims to disperse clay particles that are naturally highly stable (Albright and Wilson, 1990).

STPP is used in red and chocolate soils that are very stable and good for growing plants, but leak badly. These soils have high pore space (porosity) and require compaction to make them more impermeable. However, this is very difficult naturally, and STPP is added to the soil to complement the compaction process.

Laboratory tests are needed to determine the suitability of STPP for individual soil types because not all red soils react favourably to this treatment.

Photo 6.4 Plastic liner on the upsteam side of the dam crest.

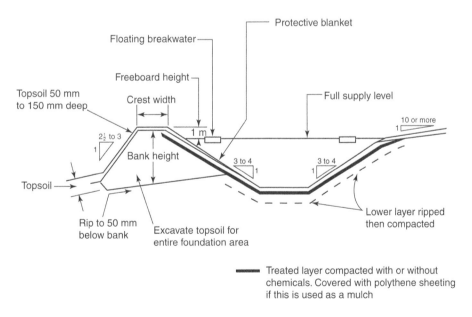

Figure 6.11 Typical cross-section of hillside dam (*Source*: SRW, 1995.)
Note: For construction in well structured friable clay soils.

Table 6.2 Cost for lining dams.

Materials	Typical material cost $/m²	Typical labour costs		Price range in total cost $/ m²
		Machinery cost $/m²	Manual labour $/m²	
Liners				
Black Polyethylene Membrane	2	–	0.5	2–5
Woven (reinforced) Polyethylene Membrane	3	–	0.5	4–8
Vinyl (PVC) Membrane	5	–	0.5	4–8
Hypalon	14	–	1	10–20
Hertalan (EPDM)	10	–	1	10–20
Butyl Rubber	12	–	1	10–20
Chlorinated Polyethylene (CPE)	10	–	1	10–20
Compacted Asphalt	7	2	5	8–16
Reinforced Concrete	10	2	6	12–24
Gunite (Air Blown Concrete)	10	20	–	10–40
Steel	10+	–	10+	20+
Compacted Soil	–	3	–	2–15
Chemical treatment				
Bentonite	7	4	1	2–10
Sodium Tripolyphosphate or Sodium Carbonate	1	2	0.5	2–20
Bitumen/Soil Mix	3	3	–	4–10
Cement Stabilised Soil	0.5	3	1	4–10
Gypsum Stabilised Soil	0.01	2	0.5	2–8
SS-13 Waterborne Dispersion	4	–	0.5	3–10
Underlays for drainage or protection				
Geotextiles	2	–	1	3–20
Sand	0.5	0.5	–	1–3

Source: SRW, 1995.

Notes:

1 Total costs of polyethylene and vinyl liners include a secondary cover of heavy duty sheeting around the top half of the dam, to protect the liner from sunlight and other causes of damage. Protective covers of soil may be applicable where gradients are no steeper than 3.5:1 (horizontal:vertical).

2 The range in total costs for bentonite, sodium tripolyphosphate, sodium carbonate and gypsum-stabilised soil includes allowance for a protective cover of untreated soil 100 mm to 1000 mm thick over the treated layers. Different thicknesses are necessary in different situations to protect treated layers from varying degrees of erosion, drying and cracking.

3 Costs do not include the provision of a firm compact base onto which liners or other treatments are placed. Dam construction costs are also additional and are valid at time of writing this book.

4 Not all types of liners or chemicals are listed in this table. Information about other treatments can be obtained from the suppliers.

5 Material costs do not include allowances for transport or delivery.

6.4.3   Artificial liners

Artificial liners are often used when the chance of natural seepage, or the cost of alternative treatments, is high (Table 6.2). Liners used to hold water can range from flexible black polythene sheet to air blown concrete (gunite).

The cost of liners varies considerably depending on site preparations, installation techniques and operator experience. While artificial liners may offer an obvious solution to a leaking dam problem (particularly in porous soils), it should be realised that most are subject to damage from stones, sticks, stock and sunlight. All lined dams should be fenced to exclude stock and other animals that might damage the liner.

6.4.4 Using a clay liner

Soil itself can be used to line a dam if it contains enough clay and if it is compacted carefully when moist. Representative samples of the clay to be used need to be tested for suitability. The blanket of clay needs to be 300 to 600 mm thick, depending on its erodability and how much it shrinks and cracks when dry. The cost of making a clay liner depends mainly on how far the clay has to be transported.

Self-installation reduces the costs of a liner or other treatment to seal a dam. There may be advantages, however, in employing people experienced in dam design, soil treatment and liner installation. Typical problems that need to be considered include:

- uplift of liners by groundwater pressures;
- uplift by wind blowing over the liner;
- damage by sunlight;
- drying, cracking, shrinking, swelling or settlement of soil;
- inadequate soil compaction;
- soil erosion; and
- damage by plant roots or animals.

Water

Two aspects of water requirements need to be considered, quantity and quality. The quantity of water required is one of the factors which determine the size of a storage (see Section 1). Other factors include:

- losses from the storage;
- practical limits on the size set by topography and materials;
- probable volumes of run-off or pumped inflow available; and
- cost limits set by available finance.

The quality of the water requirement is also important and may lead to a storage site being unsuitable for the proposed purpose. In other cases, it may be the reason why a storage is preferred to other, unsuitable sources of supply.

It is important to recognise that the quantity and quality requirements, especially for irrigation but also for stock and domestic supplies, cannot be determined absolutely. Consequently, there is no such thing as 'the water requirement for irrigated lucerne'. Both the required quantity and the limiting quality are controlled by economics. It is correct practice to use less water if it is more expensive, and to then manage with lower production in some years. The water quality which can be tolerated depends on how much water is available, what it costs, what the product is worth, and even the method of irrigation (see Section 1.9). Consumption of water by livestock is subject to considerable variation depending on the age and condition of the animal, available food, climatic conditions and water quality. Therefore, 'standard' methods of estimating water requirements, and 'standard' tables of suitable water quality provide guidelines for management and monitoring. This is particularly important in use of on-farm water supplies, particularly small dams, where the cost and quality of the water is extremely variable.

7.1 QUANTITY

One of the factors determining the planned capacity of a small dam is the potential water demand of stock, domestic services or irrigation areas. Recommended storage design procedures are based on the provision of sufficient storage capacity to meet water demands over a selected 'critical storage period'. This is a period without appreciable inflow when the reservoir must supply all water demands and losses without replenishment (see Section 1.2). Rule-of-thumb values of critical storage period are used for the design of minor storages. However, the critical storage period must be determined from an analysis of past run-off estimates for the design of major storages. In both cases, it is necessary to calculate a quantity-and-time pattern of water use for the critical period, in order to determine the storage volume needed to meet all demands and losses.

Average consumption figures may be used as a basis for preliminary planning (see Section 1.5.4). They may also be used to calculate patterns of demand for the design of minor storages, provided that these patterns are computed for the appropriate critical storage period.

Peak consumption figures should be used for the design of pumps, distribution systems and spray irrigation layouts. They should not be used for storage design, except in the case of trickle-inflow storages for irrigation use. In storage design the use of average consumption figures for estimating reservoir demand may lead to under-design, especially if the critical storage period is more than one year and includes two 'rainy' seasons. A quantity-and-time pattern of demand must therefore be calculated. This is particularly important in the design of major irrigation storages, for which a detailed monthly analysis of irrigation requirements over a critical storage period of known severity is essential (see Section 1.2.2).

7.2 WATER QUALITY

Water quality must be considered in project planning for several reasons. In the first place, the quality of the water available from a farm water system will determine the uses of that water and hence govern the overall feasibility of the system. As many natural waters have impurities that make them directly harmful to crops, a knowledge of the quality of the water supply is essential (Table 7.1). Furthermore, water quality may be one of the factors that determine the potential storage capacity. For example, if saline water is intended for irrigation use, additional quantities of fresh water must be applied periodically to avoid damage to the irrigated crops or soils. Finally, the quality of the water to be stored in a small dam must be considered in the design and construction of the impounding embankment. The presence of certain mineral constituents (including calcium, magnesium, potassium and sodium) in stored waters may be the cause of tunnelling failures in some soil types.

A variety of harmful constituents may also impair water quality and should be considered in farm water planning in terms of the intended uses (Table 7.3).

i Water for consumption by humans and animals

- Bacteria and other harmful organisms can cause illness or disease.
- Dissolved minerals can produce an unpleasant odour, or make water unpalatable.
- Silts, algae or other suspended particles can cause water to be unpalatable, discoloured or odorous.

ii Water for household, dairy or general farm use

- Dissolved minerals can cause hardness, which makes washing difficult and promotes scaling in pipelines, boilers and hot-water systems.

Table 7.1 Impurities frequently found in water.

Common name	Chemical name	Adverse effect	Appears as
Lime	Calcium Carbonate	Hardness	Dissolved Solid
Magnesia	Magnesium Carbonate	Hardness	Dissolved Solid
Gypsum	Calcium Sulphate	Hardness	Dissolved Solid
Epsom Salts	Magnesium Sulphate	Hardness & Purgative	Dissolved Solid
Glauber's Salt	Sodium Sulphate	Taste & Purgative	Dissolved Solid
Common Salt	Sodium Chloride	Taste & Purgative	Dissolved Solid
Iron	Ferrous Iron	Rust Stains	Dissolved Solid
	Ferric Iron	Rust Stains	Solid or Rust
Manganese	Manganese	Black Stains	Dissolved Solid
	Manganese Oxide	Black Stains	Black Solid
Carbon Dioxide	Carbon Dioxide	Corrosive Gas	Absorbed Gas
Rotten Egg Odour	Hydrogen Sulphide	Corrosive Gas & Bad Odour	Absorbed Gas
Marsh Gas	Methane	Inflammable Gas	Absorbed Gas
Nitrogen	Nitrogen	Inert Gas	Absorbed Gas
Oxygen	Oxygen	Oxidising & Corrosive	Absorbed Gas

Source: SCA, 1983.

- Dissolved minerals, acids or gases can corrode bore casings, pumps, pipelines and other farm infrastructure.
- Minerals or suspended impurities can discolour water and cause staining or an unpleasant odour.

iii Water for irrigation

- Dissolved minerals can inhibit plant growth or alter the character of irrigated soils.

iv Water to be stored behind small dams

- Dissolved minerals can affect the structural stability of the soils in the dam and lead to tunnelling failure.

Water samples should be tested in a laboratory to determine the presence and likely effect of impurities. Bacteria and other micro-organisms may render water unsafe for stock or human consumption. In addition, the main factors which determine the suitability of water for various farm purposes depend on the nature and quantity of the dissolved minerals or salts.

Seepage which flows both directly into the storage area and as low flows in gullies may have very high levels of salinity. This salinity is often associated with a rising water table caused by clearing of trees, or by unusually wet periods, and may well occur for the first time after a storage is built. If this saline water is allowed to accumulate in a dam during dry weather, the overall quality of the water may deteriorate so that the water is unsuitable for irrigation even after fresh run-off has occurred.

It is therefore important to check the salinity of any low flows present at the design stage. If any salinity is present, steps need to be taken to ensure that this water by-passes the storage. If this is not practicable, the site should not be used for storage.

Table 7.2 Average water consumption for farm animals.

Type of Stock	Daily consumption (Litres/head)		Annual consumption (kL/head)
	Summer	Winter	
Sheep			
ewes – lactating – dry feed	9	7	2.88
mature sheep – dry land	7	4.5	2.07
mature sheep – wet feed	3.5	2	0.99
Lambs			
fattening – dry feed	2.5	1	0.63
fattening – Irrigated pasture	1	0.5	0.31
Cattle			
grazing (<550 kg wgt)	45	30	13.5
grazing (>550 kg wgt)	67.5	45	20.7
feeding lot	94	60	27.2
calves	25	15	7.2
dairy cows in milk	70	45	20.7
cows – dry	45	30	13.5
Horses			
working	54	37	16.4
grazing	36	27	11.3
Pigs			
sows – lactating	23	18	8.2
sow & litter	45	21	–
growers (70 kg)	10	8	3.24
grower (25 kg)	5	3	–
Poultry			
laying	32 L/100	–	12
pullets or broilers	18 L/100	–	7
turkeys	54 L/100	–	20

Source: modified ANZECC, 1992 and Hislop, 1998.
Note: The above figures are only average consumption. Higher demands in hot, dry weather conditions and with dry or salty feed.

Any source of farm water other than rainwater, which is intended for human consumption should be tested to ensure that no pathogens or toxins are present.

7.2.1   Domestic use

i Problems

The main factor for determining the suitability of water for domestic uses (other than human consumption) is its hardness. Table 7.3 gives suggested hardness limits for domestic and general farm use.

Muddy water in small dams occurs when the water contains suspended matter, that is fine clays and organic material. Some of the fine clays can remain in suspension almost indefinitely. Muddy water is undesirable when used for domestic purposes. If the water is used for trickle irrigation, blockages can occur in the trickle tubes, and bacteria or mud in water supplies can cause reduction in milk quality. The treatment of muddy water is discussed in Section 7.3.

Table 7.3 Water quality guidelines for domestic drinking.

Common name of contaminant	Guideline limits (µg/L unless otherwise stated)
Biological	
Total coliforms	Should not be detected by any two consecutive 100 mL sample. 95% of samples in a year should not contain any.
Faecal coliforms	No 100 mL sample should contain any.
Algae	Up to 5000 cells/mL can be tolerated. 1000–2000 cells of cyanobacteria can cause problems.
Toxic material – inorganic	
Arsenic	50
Barium	1000
Cadmium	5
Cyanide	100
Nitrate-N	10 000
Nitrite-N	1000
Selenium	10
Organic	
Benzene	10
Carbon tetrachloride	3
1,1 – Dichloroethene	0.3
1,2 – Dichloroethane	10
Pesticides	
Polychlorinated biphenyls	0.1
Tetrachloroethene	10
2,4,5 – Trichlorophenol	1
Chemical	
Aluminium	200
Ammonia (as N)	10
Chloride	400,000
Copper	1,000
Hardness (as $CaCO_3$)	500,000
Iron	300
Manganese	100
pH	6.5–8.5
Sodium	300,000
Total dissolved solids	1,000,000

Source: modified ANZECC, 1992 and Hislop, 1998.

To minimise the risk of muddy water supplies, domestic household water for drinking and cooking purposes should be supplied by rainwater collected from the roofs of houses and sheds.

ii Precautions

Dams intended for household supplies or trickle irrigation should be fenced out from stock. Ducks and geese should also be denied access. While this is primarily a hygiene precaution, it also prevents the clay from being stirred up. It is advisable to check carefully that muddy water is not caused by erosion in the dam catchment. If this is the case, the erosion should be treated because correcting this problem is less costly than continually treating the water.

7.2.2   Stock water

General farm use is taken to include spraying, dipping, washing in dairies and milking sheds, and septic tank flushing. As a consequence of the wide variety of uses, it is not feasible to lay down specific water quality criteria.

Some properties rely on dams for their stock and domestic water supplies. They are filled from either local run-off or from water supply channels that flow only once a year to enable the dams to be filled. The main problems with this water include the high turbidity or clay colloid content and faecal pollution from animals with access to the dam. To control this, a proportion of the water supplies from the dam should be stored at regular intervals in a steel or concrete tank and treated to flocculate the clay and eliminate bacterial contamination.

The major factor determining the suitability of water for stock use is its total salinity. In general, stock can tolerate water having total soluble salts (TSS) levels far in excess of those acceptable to humans. Table 7.3 shows the generally accepted limits of TSS in water, which can be drunk by stock over long periods.

7.2.3   Irrigation water

Water can be obtained from natural run-off, diversion from streams and rivers or the development of underground resources by bore or dragline hole, but it must be of acceptable quality. In general, the quality of water for irrigation can be assessed in terms of six criteria, which together indicate the potential harm to crop, soil or equipment. These criteria are: pH, total salinity, sodium adsorption ratio, residual alkalis, specific ions, and iron (where micro-irrigation is involved). In addition, type of plant, nature of soil, extent of drainage and climatic conditions may determine the feasibility of an irrigation proposal.

7.3   WATER TREATMENT FOR HUMAN CONSUMPTION

Threshold values of bacteriological contamination are laid down by State Public Health Authorities, as are standards for turbidity, taste and odour for town water supplies. By contrast, the acceptable level of tolerance for farm water supplies relating to turbidity, taste and odour may depend upon the individual landowner.

There are several methods of treating muddy water, the most common being to use aluminium sulphate (filter alum, not common alum or potash alum) and lime, as specified below. This treatment causes the fine clay particles in the water to aggregate and sink to the bottom. However, water in the dam will only stay clear until the next run-off event, when sediments either flow into the dam or are stirred up from the dam floor. When this occurs, the water has to be treated once more. Therefore, instead of attempting to treat water in the dam, it is more effective and economical to pump a required quantity of water into a holding tank. A fully-dissolved basic solution of 1 kg each of alum and lime per 10 000 L is prepared and tested with proportionate quantities of solution and water to determine whether the concentration should be increased or reduced. If the quantities are satisfactory, the solution is stirred thoroughly into the tank water. It will take from 24 to 48 hours for the suspended clays to settle. The clean water can then be drawn off and the tank cleaned in preparation for the next batch.

7.4   ALGAE IN FARM WATER SUPPLIES

Although algae are mostly microscopic, in large numbers they can appear as floating masses or as slimes on waterways, dams, troughs, tanks or on the walls of water containers (Powling, 1990).

The major requirements for algal growth are sunlight, water, carbon dioxide and nutrients, for example, nitrogen, phosphorus or organic matter. Algal blooms can become problematic in catchments where high rates of fertiliser are applied, or where the parent rock is volcanic (basalt). In each case, high levels of phosphorus can be delivered to dams, which then provide a rich source of nutrients to promote algal growth. Most algal problems arise in late summer–early autumn when water temperatures are elevated and light levels are high. Given favourable conditions of water temperature and nutrient levels, an algal bloom can appear literally overnight.

Freshwater algae have been divided into four groups (Powling, 1990):

1   The flagellates are of all shapes, sizes and colours and are capable of independent movement by means of their whip-like projections (flagella). In large numbers, they are often responsible for the tastes and odours that can occur in drinking waters.
2   Green algae are best known by the long, green ribbons of growth often seen in rivers and channels, on the sides of tanks and drinking troughs, and in thick tangled masses in low-lying swampy areas. Green algae too may be of all shapes and sizes.
3   Diatoms consist of silica. They are usually brown or yellow in colour, and often may be seen attached to the stems of other plants, coating them with a brown slime. Other members of the group are planktonic, that is, they are suspended in the water rather than being attached to anything. They are not usually a problem in farm water supplies.
4   The blue-green algae are capable of very sudden, explosive growth and can appear as a thick green scum on the surface of a lake or dam, more particularly at the downwind end. They occur either as clumps or filaments (Powling, 1990).

7.4.1   Problems

The problems caused by algae may be physical or biological:

Physical
- clogging of filters, meters, valves and trickle irrigation lines;
- corrosion of metal tanks and other structures; and
- de-oxygenation of the water during decay of a heavy growth,

Biological

- tastes and odours, which may be particularly noticeable in water used for the house and garden;
- stock poisoning;
- death of fish and animals; and
- unsightly slimes on channels, tanks and troughs, and scums on the surface of dams.

Not enough is known yet about the mechanisms of algal poisoning, or for instance, when a scum becomes dangerous to humans or animals, although in some cases the toxin has been isolated. The toxicology of algae is very variable; research into the underlying mechanisms is ongoing. Physical symptoms in humans can vary from loss of appetite, skin sensitisation to convulsions, paralysis, and death.

Wind movement can concentrate the algal scum into a corner of the dam. If animals consume a large amount of this material, they may become very sick. However, it has been observed that these algae do not always affect animals and it is believed that only under certain conditions do the algae release a toxin (poison) into the water. A symptom of algal poisoning in diary cattle is loss of appetite and the consequent decline in the milk yield. An animal can die as rapidly as fifteen minutes after drinking.

Prevention or reduction of algal blooms requires minimising the requirements of algal growth. Although covering the water surface to lower sunlight is only practical for small dams, trees may be located to provide appropriate shade.

The bed of the dam should be cleared before it is filled to reduce the quantity of organic matter. Nutrient-rich sediments in the beds of small dams have been treated only on an experimental basis. Reeds and rushes along the dam margins will compete with algae for nutrients and will provide shelter for wildlife. Maintenance will be required to control the reeds and rushes. Cutting and removal of reeds and rushes also lowers nutrient levels (see Section 8.1.5).

Waterborne sources of nutrients include phosphorus and nitrogen in fertilisers and organic residues from stockyards, feedlots and dairies. Drainage from these areas should be diverted away from the waterway and utilised elsewhere.

If any form of algae is found in a water storage, you should still seek the advice of a person experienced in the area of water science immediately, before attempting to treat the problem.

7.4.2   Identification

The type of alga should be identified as an initial step to defining optimal treatment strategy. Identification is important if stock deaths have occurred. A sample of the alga plus water should be collected in clean drink container of one litre capacity. The container should be filled, with a small air gap and tightly sealed. A explanatory note giving details of the dam and the problem should be attached. Samples should be sent to an accredited water laboratory for identification.

7.5   SALT IN DAM WATER

Salinity is a widespread form of land degradation in many parts of Australia. Not only can soil become saline, but groundwater and surface water can be affected too.

Table 7.4 Water quality standards.

EC units	Use
0–800	• Good drinking water for humans (provided there is no organic pollution and not too much suspended clay material). • Generally good for irrigation, although above 300 EC some care must be taken, particularly with overhead sprinklers that may cause leaf scorch on some salt-sensitive plants. • Suitable for all stock.
800–2500	• Can be consumed by humans, although people prefer water in the lower half of this range, if available. • When used for irrigation, requires special management including suitable soils, good drainage and consideration of soil tolerance of plants. • Suitable for all livestock.
2500–10 000	• Not recommended for human consumption, although water up to 3000 EC could be drunk if nothing else is available. • Not normally suitable for irrigation, although water up to 6000 EC can be used on very salt tolerant crops with special management techniques. Over 6000 EC, occasional emergency irrigation may be possible with care or if sufficient low salinity water is available, this could be mixed with the high salinity water to obtain an acceptable supply. • When used for drinking water by poultry and pigs, the salinity should be limited to about 6000 EC. Most other livestock can use water upto 10 000 EC.
Over 10,000	• Not suitable for human consumption or irrigation. • Not suitable for poultry, pigs or any lactating animals, but beef cattle can use water to 17 000 EC and adult sheep on dry feed can tolerate 23 000 EC. However, it is possible that water below these levels could contain unacceptable concentrations of particular ions. Detailed chemical analysis should therefore be considered before using high salinity water for stock. • Water up to 50 000 EC (the salinity of the sea) can be used to flush toilets provided corrosion in the cistern can be controlled, and for making concrete provided the reinforcement is well covered.

Source: modified from SR&WSC, 1970.
Note:
EC units means electrical conductivity, in micro-siemens per centimetre (μS/cm).
To find the approximate mass of salt in milligrams of salts per litre of water (or kilograms per megalitre of water) multiply the EC unit by 0.6; e.g. water of 2500 EC has 1.5 tons of salt per megalitre of water. A megalitre of water has a mass of 1000 tonnes.

This section outlines ways of minimising the chances of water in your dam becoming salt-affected. If your dam has already become salty, a ready reckoner for measuring the suitability of the water for stock use is provided (see Table 7.4).

With the intense development of farms and associated difficulties in finding suitable dam sites, poor quality (saline) water is increasingly likely to be found. There is no point in building a dam which will collect saline water, particularly as high rates of evaporation may increase the concentration of salt in the storage. Before building a dam, check for saline ground in the catchment area along drainage lines, around the site and in areas where the water from the catchment gathers or disperses.

Under certain conditions, it is possible to design works in dams which enable saline surface and sub-surface flows to be intercepted and passed through or around the dam before affecting the water in storage. These works are ideally included in the original dam design but can be added at a later stage. Techniques for diverting saline water away from dams are outlined in Section 7.5.1.

Where salinity threatens to affect an existing dam, the vegetation cover should be increased within the catchment by planting trees and deep-rooted perennials. This will help to minimise recharge to the groundwater system. High groundwater flows can upset the existing salt balance by flushing large quantities of soluble salts, previously stored in the soil profile, to dams through the stream system or directly from the underlying soil. Water from over-irrigation and residual salts from fertilisers, may also contribute to the problem.

7.5.1 Minimising salinity in dams

Surface interception techniques to improve water quality are suitable for protecting new works or when installing an existing storage. Surface water interception relies on diversion works being installed in a waterway or gully upstream of the dam. Flows of saline surface water are intercepted either by a suitably protected low embankment or diversion drain. The drain needs to be constructed slightly off the contour to carry flows around the storage to be discharged into the waterway downstream of the embankment or an adjacent gully.

A limitation of this technique is that the diversion is not effective for permeable soils where flows will eventually reach and pollute the storage. This can be overcome by collecting the low flows in a sump and piping them either around or directly through the storage, before discharging into the drainage line. Where a diversion pipe is used and is laid around the storage, the available head will be limited. Care will be needed when installing the pipe to maintain a uniform grade and to prevent the formation of airlocks in the line.

7.5.2 Interception of sub-surface saline water

Sub-surface interception works involve a cut-off trench below ground level. This extends into permeable material to intercept saline flows. Upstream of, and adjacent to this cut-off, a filter drain helps to collect the flow and deliver it into a diversion pipe. This pipe passes either around or through the storage. Since the diversion pipe functions as a siphon, it must be periodically primed before operation.

When sub-surface saline flows are found in your catchment and you are considering the most suitable technique for interception, further investigation should be carried out to determine the source of these flows. If a particular gully or hillside is identified as the source, diversion works should be installed in the specific area. However, if no point source can be determined, it may be necessary to consider an installation beneath the storage.

Interception beneath the storage has the advantage of requiring a much shorter diversion pipe. It also allows the main cut-off to perform the dual function of preventing seepage beneath the embankment, and intercepting saline flows. However, this technique should be viewed with caution, as there is a possibility of the saline flows mixing with water from the storage before reaching the interception works. An ideal situation exists where (a) the bed of the waterway is covered by a clay layer that overlies rock, and (b) where salt inflow from the surrounding topography occurs above the rock. The clay membrane helps to suppress the poor quality flow, which is then removed by the diversion.

7.5.3 Maintenance

Maintenance of sub-surface and surface interception works is essential for their continual successful operation. The long-term reliability of these works is particularly important when it is necessary to reprime siphons or to periodically check flows. There is a real danger that maintenance can be overlooked if the farm changes ownership, particularly where the incoming landowner or share farmer may not fully understand the significance of maintaining the works in operating condition.

7.5.4 Downstream effects

Consideration should be given to the discharge of poor quality water downstream of the dam or storage. In most cases, this would be the natural destination of the water if the storage had not been constructed. The installation of interception works does not change the situation with respect to downstream properties. However, as the saline flow will have been diverted to the surface by the works, you must consider your common law obligation or 'duty of care' (see Section 10.2) not to harm another landowner. You may also need to consider and obtain permits under other local legal requirements such as Proclaimed Catchments or Planning Schemes (Local Government). If, for some reason, you are unable to discharge to a waterway, then a natural or artificial evaporation basin should be considered.

7.5.5 Minimising the effects of evaporation

Evaporation concentrates salt in water. The surface area of the stored water and local climatic conditions determine the evaporation rates. A narrow, deep storage will lose less water by evaporation than a broad, shallow storage.

When building a new dam it is important to make it a minimum of 2 m deep to avoid it becoming a holding basin for salt every time the water evaporates. Constant evaporation leads to a build-up of salt on the base of the storage. Each time new water flows into the dam, this salty base dissolves into the clear water, polluting it with salt. To minimise this effect where possible, skim off the top 100 mm of salt residue using a grader blade. Remember to deposit the resulting sludge in a place where it cannot leach back into the storage.

7.5.6 Stock and salty water

Where water is salty, as is often the case with groundwater, you may have difficulty deciding whether your stock can drink it without harmful effects. Although cattle and sheep can survive on quite salty water, better quality water is required for poultry and

Table 7.5 Range of total saline matter in water supply.

Human Livestock Horticulture	Preferred maximum (ppm)			Upper limit dry conditions (ppm)		
	mS/cm	EC	ppm	mS/cm	EC	ppm
Drinking (human)	0.83	830	500	2.5	2500	1500
Biological effects	–	–	–	1.3	1300	780
Fruit trees	–	–	–	3.0	3000	1800
Poultry	3.3	3300	2000	6.7	6700	4000
Pigs	3.3	3300	2000	7.5	7500	4500
Ewes – lambing	5.0	5000	3000	10.0	10 000	6000
Dairy cows	5.0	5000	3000	10.0	10 000	6000
Horses	6.7	6700	4000	11.7	11 700	7000
Cattle graze	6.7	6700	4000	16.7	16 700	10 000
Sheep (wet feed)	6.7	6700	4000	20.0	20 000	12 000
Sheep (dry feed)	8.3	8300	5000	25.0	25 000	15 000
Yabbies	23.3	2330	14 000	41.7	41 700	25 000
Pacific Ocean	–	–	–	58.3	58 300	35 000
Dead Sea	–	–	–	550	550 000	330 000

Source: modified SR&WSC, 1970.
Relationship between units: 1 dS/m = 1 mS/cm = 1000 EC = 1000 μS/cm = 600 mg/l = 600 ppm.

pigs, and for animals that are lactating or growing. The suitability of water for stock is usually assessed in terms of the concentration and composition of dissolved salts. The limits adopted Australia wide for total soluble salts in drinking water for stock are shown in Tables 7.4 and 7.5.

Section 8

Ecology

8.1 WILDLIFE AND PLANTS IN DAMS

Most small dams are designed and constructed primarily for the conservation of water for stock, domestic and irrigation purposes. However, while a dam can seldom be devoted wholly to the conservation of wildlife, many farmers regard this as an important secondary use.

If attracting wildlife is desired, the actual siting and shape of the dam are important considerations. Unfortunately, the basic rules of design relating to stock, domestic and irrigation dams vary in some ways from that of an ideal wildlife dam. Furthermore, the characteristics of a dam designed for fish will vary from one designed for waterbirds. However, with careful planning it is generally possible to reconcile the various uses required of the dam and to modify the design, siting and construction of most dams to suit all purposes. In most Australian States, the wildlife associated with small dams is almost exclusively waterbirds, with other forms using the margins if these are suitable. Waterbirds, and particularly wild ducks, are attracted to almost any stretch of water, whether it be natural wetland or land deliberately flooded for a specific purpose. However, birds will only remain if these water areas provide food, shelter, breeding sites or other basic necessities.

The key to food production for waterbirds is water, and most areas of wetland which are dry for a part of the year and then flooded, can produce bird food as a regular part of the soil and water conservation plan for a farm. The level of this supply depends upon a number of factors, including quality of the soil, vegetation type, flooding frequency and depth, and present climatic conditions.

In general, small dams require a certain aesthetic value to attract waterbirds. They do not support a waterbird population in proportion to their area, nor do they provide suitable habitat throughout the year. There are exceptions to this, but in most places the importance of small dams to waterbirds is a reflection of the huge number of small dams in existence rather than the quality of habitat provided by them.

These dams could make a very real contribution to waterfowl conservation if made into a more suitable habitat. In addition, small dams can provide very important habitat niches to frogs, which are an essential component of healthy ecosystems.

Irrigation schemes can benefit wildlife in many ways, particularly in the drier areas of the country, by increasing the food supply and sometimes the amount of cover. Food sources include unharvested grain and insects in irrigated pasture. Furthermore, the waterways and drainage pondages are sites for breeding, feeding and resting. Balanced against this is the drainage of natural swampland, and the clearing and channelling of waterways, all of which decrease or fragment habitats for wildlife. Any wildlife management program in irrigated regions must provide adequate breeding facilities, which are associated with an ample food supply as well as shelter. Government agencies are endeavouring to provide this on State-owned land but the assistance of the general farming community is necessary if the land is to support sufficient numbers of waterbirds.

If the present waterfowl populations are to be maintained, we must not only develop existing wetlands, but create new areas of suitable habitat. The characteristics of a small dam, which are relevant to wildfowl conservation, are discussed in Section 8.1.7. It should be noted that many of these characteristics conflict with the requirements for fish production, and it is often difficult to reconcile these in the management of any one dam. However, dams of both types occur, and there is a sufficient number to allow management for both fish and wildlife conservation.

8.1.1 Water regime

The amount of wildlife attracted to a small dam is determined by the area of the water surface. A surface area between 0.2–1.0 ha has been found to be the most practical. The feeding depth for most ducks and many of the wading birds in eastern Australia is 25 to 300 mm, and therefore a large area of shallow water is required. This is often only provided when a small dam overflows, but could well be a feature in the construction of new dams. Periodic drying out and flooding of this shallow water environment, with the subsequent growth of rank, ungrazed grasses and herbs provide ideal conditions for the maximum production of duck and more importantly, duckling food.

8.1.2 Basic topography

The edges of dams should be long and shallow, and retain as much topsoil and vegetation as possible in an undisturbed condition. In addition, the ideal waterfowl pondage should have a shelving margin. Waterbirds prefer to walk out of the water to feed, rest, or reach a nesting site rather than to fly onto dry land. The edge of the dam should be as irregular as possible, as the maximum use of the margins is made by waterbirds for feeding and resting. Consequently, the longer margin provided by an irregular edge to a dam may outweigh considerations of the acreage of the dam.

A desirable feature of any waterbird pondage is the inclusion of a number of islands, which provide excellent breeding and resting sites, even if they have relatively steep edges. Islands afford protection from disturbance, particularly from predators in the breeding season. Half-submerged logs, mud banks, or open flat margins can provide other useful nesting sites.

8.1.3 Vegetation

Trees and shrubs can be used around small dams to attract wildlife, especially birds. To be fully effective, planting for habitat should be done in combination with the provision of feeding and shelter sites. Shelterbelts near dams can also provide shade for stock and reduce evaporation by lowering the wind speed at ground level and decreasing surface temperatures. In addition, frog species favour dam sites characterised by a high percentage of emergent vegetation in the dam, low levels of bare ground in the immediately adjacent area, and a high percentage of tree cover within an approximate 200 m radius of the dam site (Hazell *et al.*, 2001).

Trees and shrubs should be placed carefully to avoid problems occurring in the future. Trees should never be planted on the dam embankment and spillway, or near the outlets. Trees, shrubs and grasses should be established around the margins, and up to the waterline, using native species wherever possible. River red gum (*E. camaldulensis*) can be planted in shallow water or on land subject to flooding, to provide shade, and eventually nesting habitat and shelter for some species of waterbirds. However, care should be exercised in the selection and placement of tree and shrub species. Too many tall trees around a small dam will make it unattractive to waterbirds, particularly if these are across their normal flight lane on and off the water. Low shrubs will afford birds better cover and minimise interference with their flight pattern, although a few tall trees are still needed.

The planting of food crops for waterfowl has proved successful overseas, but there has been little investigation into this aspect of waterfowl management in Australia (Cowling, 1967). Anecdotal evidence suggests that sufficient food sources are provided from pasture and grain crops and by the flooding of most native vegetation.

8.1.4 Grazing by stock

Grazing the margins and shallow water zones of small dams reduces food supply through trampling and manuring, and is therefore detrimental to the provision of waterbird habitat. The exclusion of stock by fencing the dam, including a reasonable dryland margin, allows the area to provide sufficient food, shelter and nesting sites. Many species of waterbirds nest in grass or other ground vegetation. Stock do not need to be excluded for the entire year. An alternative means of watering, or a fenced lane can be provided.

8.1.5 Waterways and swamps

Small dams in irrigation areas have the potential to support whole populations of waterbirds. Under these conditions, there is a significant acreage of waterways (for both irrigation supply and drainage), and swamps which are too low to drain, salty or receive drainage water. These areas can also be developed to produce wildlife. In particular, those pockets of unusable land along creeks, or low-lying swamps, can be fenced and planted with native vegetation, or even cultivated and sown with grass or food crops.

A number of the more important duck foods are relatively salt tolerant, and salty land can be converted to waterfowl feeding areas by flooding with excess irrigation water, or drainage water from irrigation. In times of flood, water often reaches these lower, saltier areas with comparative ease. Often these areas can be planted with red gum and used to provide shelter for stock, or rough grazing if the farmer does not wish to devote the entire area to wildlife production. While these areas may be small in relation to larger and more spectacular natural swamps, they would be large in relation to the more permanent water of the small dams.

Often swamps carry a dense growth of cane-reed, cumbungi or rushes (see Section 8.2.1). These provide ideal shelter, but little food, and often waterbirds are deprived of sufficient open water to land or take off. These areas should be partly cleared to provide sufficient landing space and food production, but some non-invasive reed should be retained for shelter. Research currently underway indicates that small dams offer a significant array of niches for wildlife that may maintain certain species in otherwise hostile environments. Dams appear to provide a chain or corridor of opportunities. Hunting within this setting is not an option.

8.1.6 The role of hunting

Currently, pressures for public recreation are being exerted from intensively developed urban areas, and from the rural population. However, the effect of this additional pressure on natural resources has been disastrous for many species of wildlife. Hunting is one of many pressures acting upon wildlife populations, and often it accelerates, or even completes, a process of extermination of a species that commenced with environmental change. These changes and associated impacts include:

- drainage of breeding areas;
- competition with introduced animals for food;
- clearing;
- the insidious effects of pesticides or effluent; and
- altering predator–prey relationships.

Hunting reduces wildlife resources immediately, but there are also longer-term impacts imposed by loss of biodiversity and vulnerability in reduced, fragmented populations.

Hunting pressure on a small dam will drive wildlife to less disturbed habitats, which may not be as suitable for sustaining populations. This is particularly important given the tendency for some duck species to stay close to the dams where they hatched, and their preference of dam sites within 400 m of tree hollows (Kingsford, 1992). The response to seek less disturbed habitats under these conditions may also have longer-term impacts on ecosystems since the interactions between species will be modified. Therefore, it is imperative that hunting on or near small dams should not be practiced.

8.1.7 Waterfowl management

The basic points for siting and construction (Cowling, 1967), development and maintenance of small dams for wildlife purposes include:

i Siting and construction

- Ensure that there is sufficient shallow water to produce food and allow birds to feed.
- Provide at least one island as a resting and breeding site.
- Endeavour to provide the maximum amount of edge by an irregular margin.

- Provide a reasonable area of shelving edge to allow birds access to dry land.
- Place old logs, beams, in the shallow water as resting sites.
- Ensure adequate stabilisation of banks and waterways.

ii Development and maintenance

- Fence at least part of the water's edge to exclude stock, particularly the shallower parts of the dam.
- Remove any vermin and noxious weeds from around the dam. (These cut down the production and availability of food.)
- Ensure that ground vegetation will develop, either naturally or by cultivation and sowing.
- Plant the margins with native trees and shrubs, limiting tall growing species to a few trees.
- If necessary, provide artificial nesting sites in the form of 'rafts' or nest boxes. Remember that birds will not breed if there is no food for themselves, or the young they will produce, regardless of the quality of nesting.
- Regulate the hunting pressure to a moderate level to retain the attractiveness of the area for waterfowl.

8.2 WATER PLANTS IN DAMS

Aquatic plants in small dams can block pump and pipe inlets, deter stock from drinking, and in some cases, taint the water. If plants are treated when they first appear, the dam can be kept relatively free of the more troublesome species. Dense weed growth or outbreaks usually indicate that the water is high in nutrients.

All plants can become a problem if allowed to spread and disrupt water flow, and each species may require a different control method. When considering control measures for aquatic plant problems, it is recommended that the following steps be taken:

- correctly identify the plant, using a good reference guide or by consulting your local naturalist or botanist.
- decide whether you will use mechanical methods (such as raking, slashing, mowing, grazing and burning) or chemical methods, or a combination of both.
- if using chemicals, use only the recommended quantities and follow all safety precautions listed on the container labels. Remember that greater quantities of herbicide do not produce better results.
- avoid spray drift by spraying only on calm days.
- use of chemicals should be avoided if possible in or near waterways or drains flowing into rivers or creeks.

8.2.1 Aquatic plants

The aquatic plants that are best known to most of us are the varieties of water lilies we see either in urban lakes and small dams or in fish ponds in home gardens, and eel grass associated with aquariums. These plants rarely cause problems.

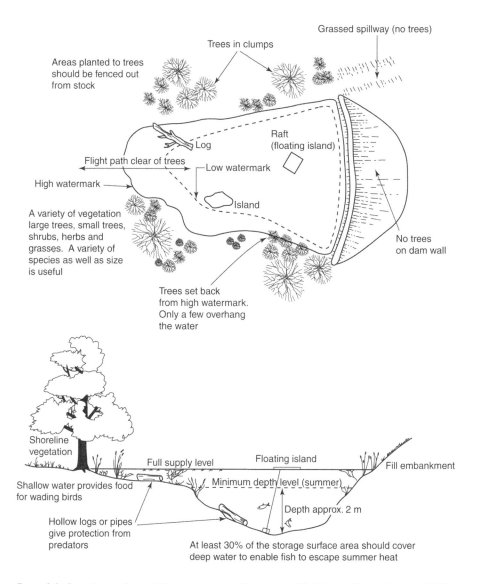

Figure 8.1 Dam layout for wildlife conservation (*Source:* modified from Hill and Edquist, 1982).

In the management of water resources in Australia, the aquatic plants of most concern are duckweeds (*Wolffia* spp.) and azollas (*Azolla filiculoides*), arrowhead (*Sagittaria montevidensis*), cane grass (*Eragrostic australasica*) and cumbungi (*Tyha domingensis*), alligator weed (*Alternanthera philoxeroides*),water hyacinth (*Eichhornia crassipes*), salvinia (*Salvinia molesta*) and lagarosiphon (*Lagarosiphon major*). These plants have been declared noxious throughout Australia because of their potential to disrupt the proper utilisation of water resources through the blocking of streams and irrigation channels. Being noxious, these plants must be destroyed wherever they occur.

These species have the potential to cause problems in water supplies. An infestation that remains unrecognised for several years will spread and become increasingly difficult or impossible to eradicate. Consequently, early recognition of a potential problem is essential.

Reporting the occurrence of these species to the relevant State agency or water authority is recommended even when correct identification of the plant may be in doubt.

a Duckweed and azolla

Duckweed and azolla are free-floating aquatic weeds that grow and flourish on slow-moving or ponded water, including small dams and irrigation channels. These plants can cover the water surface completely. As a result they may block pump-inlets and deter stock from drinking the water. Both weeds use nutrients in the water for their growth, and in most cases, wash-off from heavily stocked or over-fertilised areas of the farm contribute to the problem.

The two weeds are easy to identify. Duckweed has round green leaves about half the size of a match head. Azolla has slightly larger fern-like leaves, which are usually green in their early stage or when growing in shaded areas, but redden on maturity or in full sunlight.

These plants should not be confused with algae, which are fibrous like hemp or cotton wool, or appear as scum on the water surface.

Duckweed is commonly spread by wild ducks carrying the seed on their feathers from dam to dam, or by flushing downstream during heavy stream flows.

Duckweed and azolla are best controlled by:

- lowering the nutrient level in the water, by stopping inflows to a dam from heavily stocked or fertilised areas;
- re-filling the dam, as often as possible, with good quality water; and
- cleaning out silt from the bottom to remove built-up organic matter.

In case of dense growth, as much of the weed as possible should be removed by raking or 'skimming' it off the top of the water, and disposing of it appropriately. This job is easier when there is a strong wind that helps blow the weeds to one side of the dam. Raking reduces the amount of herbicide required to kill the plants, and helps reduce the likelihood of an algal bloom caused by the breakdown and decay of sprayed plants. Any remaining weed can be treated with herbicides available from agricultural chemical suppliers, stock and station agents and cooperative trading stores.

After spraying, water from the dam should not be used for domestic purposes or sprinkler irrigation for at least 10 days, or for stock for at least a day. This precaution is necessary as most chemicals take some time to break down in water. However, they may not be harmful to fish or wildlife. It is recommended that safety precautions listed on the container label are closely observed when using any farm chemical. Follow-up treatments are also recommended to prevent further infestations.

The entire dam does not need to be treated with herbicides as the plants will settle to the bottom of the dam and decay, releasing more nutrients to continue the cycle of plant and algal growth. If these treatments are unsuccessful, it may be necessary to empty the entire dam, clean out the plants, remove silt, and re-fill with clean water. This will help remove the nutrients that contribute to plant growth.

b Arrowhead

Arrowhead is becoming increasingly prevalent in northern Victoria. Although it has not spread as extensively as some plants, it has the ability to spread rapidly, resulting in restriction to the flow of channels, drains and creeks. Arrowhead is an erect plant which grows up to 1 m high. Its leaves are arrow shaped and are a distinctive dull-green colour. Its flowers are white with yellow centres carried on fleshy stems.

This plant is commonly found growing in shallow, slow-flowing creeks, channels and drains. It spreads rapidly, forming dense clumps which slow water flow significantly. Arrowhead reproduces by seed germination, underground rhizomes (horizontal stems that put out roots and shoots), and bulbs which remain viable in the soil for many years. Due to these various methods of reproduction, it can be difficult to control this plant mechanically. Any rhizome or root fragments left in the soil will result in a recurrence of the infestation.

Herbicides appear to be the only effective way to control this plant. Repeated treatments are usually necessary because of the long germination period from September to December. For herbicide rates, contact your nearest chemical supplier.

c Cane grass and cumbungi

Cane grass and cumbungi (bulrush) can usefully provide a habitat for fish and wildlife, and help prevent riverbank erosion. However, if these plants are allowed to spread across a stream, water flow may be restricted and the banks may deteriorate, leading to flooding on either side. You must be careful when controlling these plants that you do not completely denude the area.

Control measures include the use of a dragline in combination with an appropriate herbicide. Using a dragline to pull out cumbungi and cane grass gives immediate results but is often impractical because of poor tractor access, and weed regrowth from broken rhizomes and roots. If weeds need to be cleared immediately, the use of a herbicide followed by excavation with a dragline, no later than two months after spraying, is recommended.

Where access along banks or embankments is difficult, spraying from a small boat may be feasible. Chemical control used on its own does not result in immediate clearance of the weeds, as they can take several years to decay completely. In the case of dense, mature infestations, burning old foliage before the new season's growth starts is an effective way of reducing the amount of plant matter. Ask your chemical supplier to recommend an effective herbicide.

d Alligator weed

This perennial creeper has shiny, green leaves and white clover-like flowers. It forms dense, floating or rooted mats up to 1 m depth, and also thrives on dry land sites. It spreads by stem fragmentation.

Alligator weed is arguably the most destructive aquatic plant in Australia. At present, it covers several thousand hectares of coastal New South Wales. Biological control has been successful only in isolated situations. Overall there is no control in river-banks or on other dry land sites.

e Water hyacinth

Water hyacinth continues to be a serious threat to waterways and storages throughout Australia, but control measures and inspection programs undertaken by State authorities have prevented any significant increase in its distribution in recent years. It is a free-floating perennial plant that grows to almost 1 m in height. The leaves have distinctive swollen stems and the flowers are mauve in colour. The plant spreads both by seed and by stem fragments.

All infestations that have developed in some States to date have been eradicated. Herbicides have been used as a control measure in a number of situations and biological control agents have led to a substantial reduction in levels of infestation in Queensland and to a lesser extent in northern coastal New South Wales.

f Salvinia

Salvinia is a threat to Australian waters which rivals water hyacinth. At some locations, where water hyacinth has been brought under control, salvinia has replaced it. Salvinia is a free-floating and hairy aquatic fern. It spreads by fragmentation and grows rapidly, the area of infestation doubling in two to three days under warm conditions.

Herbicides have been used to control some infestations of salvinia. Biological control agents have proved successful at some locations in the Northern Territory and Queensland.

g Lagarosiphon

Lagarosiphon has the capability to cause major problems in water storages and waterways, particularly in southern Australia. It has been found amongst plants being sold in the aquarium trade. Lagarosiphon is a submerged perennial with narrow, recurved leaves on branched stems up to 5 m long. It prefers cool conditions and grows well in water with low nutrient levels on a silty or sandy bed.

8.3 USING HERBICIDES NEAR WATER

It is important to remember that the landowner is legally responsible for injury or damage caused by the careless use of herbicides (see Section 10). Users of herbicides have the responsibility to select the most suitable herbicide for the intended purpose and to use it in a way that does not have adverse effects.

Herbicides are chemicals that have been developed for the purpose of killing plants or controlling their growth. When herbicides are used near waterways, storages and other water bodies, several aspects must be considered that are of special significance. Only a few herbicides are effective against plants that grow in or near water. These herbicides are toxic to some extent, and consequently, care must be taken to ensure that their use does not cause injury to people, stock, fish and wildlife, or damage crops and other useful plants.

Amongst the more important requirements are the correct rate and time of application, and depending on the particular herbicide being used, whether it is applied to the foliage or the soil. To ensure maximum effectiveness, herbicides must be applied in the way recommended on the labels. Where submerged species require treatment, the herbicide may contaminate the water.

8.3.1 Hazards

Herbicide residues must be considered in relation to the use made of the water. There are approved residue limits for most of the herbicides used near water, and for a particular herbicide, the limit may vary depending on whether the water is used for domestic purposes, agriculture or recreation, or if fish or their habitat are of special significance to the area. In addition to choosing the herbicide that will achieve the desired result, the user must ensure that it is applied in such a way that the appropriate residue limit is not exceeded (Bill, 1985).

The degree of risk from herbicide residues in the water is determined by the toxicity of the herbicide and by the concentration and persistence of herbicide residues in the water. Plant decomposition following the use of a herbicide, and de-oxygenation of the water which kills fish are indirect impacts of herbicide use. Therefore, it is essential that the user be aware of these considerations.

Important points that should be observed when a herbicide is used near water are summarised below.

- Use only a herbicide that is approved for the purpose and apply as directed on the label and in accordance with conditions specified by the approving authority.
- Identify the uses made of the water and ensure that the proposed herbicide treatment does not affect those uses adversely, and that any residues of herbicides in the water do not exceed approved limits.
- Where the water is used by the public, obtain approval for the proposed herbicide treatment and notify people nearby of the nature of the treatment and any restrictions on the use of the water.

8.4 VEGETATION ON AND AROUND DAMS

Clumps of tall trees that shade the water's edge in summer are important in providing shade and shelter (see Section 8.1.3).

The quantity of water used by trees growing at a dam edge will vary greatly depending on factors including the size, number, location and species of the trees, and the soil and weather conditions. Water use by trees at a dam edge is not usually a problem where a dam is large, the ratio between catchment size and dam size is favourable, or rainfall is high and reliable. Even with small dams, or when rainfall is less reliable, nearby trees can still provide overall benefits providing they are not established too close to the dam edge.

8.5 YABBIES

There are many species of yabby or freshwater crayfish that can damage dam walls by burrowing through the clay. Although it is commonly believed that yabbies are responsible for damage to dam walls, they should not be blamed automatically for all leaks or damage. Tree roots may also weaken the banks of dams, and yabby holes seen outside the water are probably due to land yabbies, which are different from the aquatic species. Increasingly, people are regarding yabbies as a recreational and food resource and are glad to have them in their dam. If you are sure aquatic yabbies are causing damage, you can take several steps to reduce their numbers.

8.5.1 Physical removal

Constant yabbying may reduce the numbers sufficiently to prevent damage. 'Dragging' is probably the quickest method. This involves throwing a few pieces of red meat into the dam to draw the yabbies out of their holes, then, with the help of another person on the other side of the dam, slowly dragging a net through the water.

8.5.2 Biological removal

Yabbies form part of the diet of some native fish. Stocking a dam with fish may help reduce the numbers of yabbies but is unlikely to completely eradicate them. Stocking with larger fish is preferable, as they are immediately physically capable of eating the yabbies. Furthermore, their survival rate is proportionally higher than smaller fish and as such likely to be ultimately more cost efficient. The number of fish required is determined by the size of the fish and the area of water surface, rather than the volume of water in the dam.

The best fish to use for reducing yabby numbers are golden perch (also called callop or yellow belly), silver perch, or Murray cod. These fish are readily available commercially. Redfin are no longer recommended as they have been found to carry a disease harmful to trout, and their tendency to over-breed results in too many small fish. However, they have a great preference for yabbies and can tolerate a wide range of environmental conditions.

Exotic fish species such as brown and rainbow trout are also suitable, although conditions in small dams are usually unsuitable for trout to spawn. This means you will have to re-stock at regular intervals to maintain the fish population. Trout have a relatively fast growth rate and are a good sporting fish, but conditions for their survival are critical. They need clear, cool water with a summer temperature not exceeding 19°C, and at least one-third of the dam should be 2 m deep.

When only fry are available, a stocking rate of about 2500 per hectare is necessary to allow for their expected high mortality. The stocking rate recommended for yearling-size perch is about 250 to 350 per hectare.

More fish can be supported when the water quality, natural cover and vegetation are all in good condition. However, for larger fish the stocking rate is lower. An adequate natural food supply is needed to enable the fish to grow to a size where they can eat the larger yabbies. It is therefore helpful to allow some aquatic plants to grow around the water's edge which, in turn, encourages small invertebrates, such as mud eye, on which fish feed. The survival rate of fish will be higher if a dam is managed for fish-eating birds such as cormorants.

It is important to remember that a permit is required from a State agency office before purchasing the fish.

8.5.3   Chemical control

Chemical methods of eradicating yabbies, including manipulating water pH and adding chlorine, are no longer recommended because of possible long-term damage to aquatic ecosystems, and because they are not proven to be effective control measures.

Commercial

Some of the information provided in this section is from *Notes* and *Infosheets* from the Department of Conservation and Environment, Victoria, Fisheries Management Division (DC&E, 1990).

This Section of the book is not intended as a complete guide to freshwater aquaculture or investment in commercial aquaculture, but provides a broad overview. A commercial venture into aquaculture deals with the same basic principles, and even some of the same species as small-scale operations. The design of a pond or a dam for commercial production must be different in large-scale enterprises for ease of maintenance and harvesting.

Aquaculture is now an important industry with increased future export potential (Lewis and Branson, 1996). There are a variety of definitions of aquaculture, but for the purpose of this book it has been defined as the activity of husbandry of aquatic organisms. Most commonly this is regarded as the value-adding to fish, shellfish and plants living in water. Generally, few species of fish have been found suitable for farming and, in Australia, most fish farming enterprises are in the early stages of development with an associated moderate economic risk.

Inland salmonid production, which includes rainbow trout and some Atlantic salmon, produces the most economic wealth for some Australian States. Production from this type of aquaculture has steadily increased over the last few years. The second largest aquaculture sector is aquarium fish. This consists of still water goldfish and the on-growing of imported stock in quarantine rooms. Eel production is the next most common form of aquaculture. Over the last ten years this form of aquaculture has gradually increased as the result of improved technologies. Inland yabby farming is another popular form of aquaculture. However, although potential for profitability exists, yields have generally not been high and production has been fairly static over recent years. Finally, there is some production of the fry of native species such as silver perch and Murray cod. Although not commercially produced at present, inland barramundi production shows some potential.

The attitude of the public to aquaculture tends to vary between optimism and scepticism. Aquaculture is a moderately risky form of enterprise that is capable of fair but sustained growth if undertaken sensibly. Overseas experience has shown that poor planning and management of the aquaculture industry leads to a boom or bust cycle and associated environmental degradation. Freshwater aquaculture developments are non-consumptive users of water because it is returned to the waterway after use. However, the discharge from these enterprises contains nutrient-rich effluent which has water quality implications for waterways.

9.1   FISH FARMING

In theory, there are many species of fish that could be farmed commercially, but few have been found to be suitable. Most species require particular sets of water conditions, which are affected by climate, and are therefore often best suited to specific areas.

Before deciding on a species to grow, it is important to first examine existing ventures in the proposed area. If your choice of fish is not currently being farmed, find out why, as it is probable that there are good reasons to avoid this type of fish.

9.1.1   Established freshwater species

Today there are, at least, three freshwater species that have formed the basis of a successful aquaculture in Australia with a fourth showing some potential. The farming of rainbow trout is a proven success of long standing. However, most of the more suitable sites are occupied and there is little room for expansion. For a commercial farm, a large quantity of high quality, cool water is needed every day of the year and there are few places remaining where this is available.

Eels are successfully and extensively farmed. Natural waters are stocked and harvested using traditional commercial fishing methods. Most of the suitable waters are now committed and there are possible limits on the supply of seed stock. The opportunity for new enterprises is limited. There may be potential for intensive eel farming on private land but there are many problems that have to be solved before this method is viable.

Goldfish are grown for supply to the aquarium trade. Ventures in this area have been amongst the most financially rewarding forms of fish farming, but the market is

Figure 9.1 Oncorhynchus mykiss (Rainbow Trout) (*Source:* DC&E, 1990).

Figure 9.2 Anguilla australis (Short Finned Eel) (*Source:* DC&E, 1990).

now probably close to saturation. However, there may be room for limited expansion in production of high quality, fancy strains.

9.1.2   Dam conditions that control productivity

The major factors in determining which species of fish are best suited to a dam and what level of productivity can be expected include:

- size, shape and depth of the dam;
- the depth of water remaining during the summer;
- the amount of overflow; and
- the water quality and temperature.

In general terms, the ecosystem of a dam is dependent on the following conditions:

- productivity, which is usually more dependent on surface area than volume; and
- depth, which provides refuge from predation and cooler temperatures for species. However, depth does not contribute directly to the productivity of the dam.

Natural food supplies for fish within a dam come from two basic sources:

- terrestrial animals that find their way into the dam and are eaten by the fish; and
- aquatic organisms that are part of the basic food chain within the dam.

The basis of the food chain in a dam is algae, which utilise available sunlight and nutrients, such as phosphorus and nitrogen, to grow and breed. These algae are consumed by other organisms in the dam that are in turn consumed by the fish. The aim of sound aquaculture practices is to balance the numbers and size of fish in a dam to optimise productivity. Unwanted algal growths, which can produce low dissolved oxygen levels, can result from excessive nutrient inputs from agricultural fertilisers, run-off from dairies or piggeries, or natural run-off from basaltic catchments. High nutrient levels will ultimately lead to an insufficient food supply, reduced growth rates

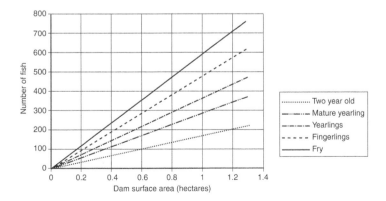

Figure 9.3 Stocking rate for trout (*Source:* DC&E, 1990; Hill *et al.*, 1982).

and poor fish survival. High stocking rates will also result in poor growth rates and poor survival.

Apart from embankment failure or loss of water through leakage, the biggest threat to any fish dam is input of sediment. Erosion in the catchment from gullies, over-cultivated or over-grazed fields and poorly vegetated hillslopes/valley bottoms are all sources of sediments that are delivered to a dam. When sediments flow into a dam, they settle to the bottom, thereby decreasing its depth and capacity over time. In addition, sediment build-up within a dam may produce increased pressure on the spillway, increased water temperatures in summer and decreased productivity of the fish farm.

Some operators of fish farms suggest that trout require at least one-third of the dam to be at least 2 m deep during summer, whilst native fish can survive when at least one-third of the dam is 1.5 m deep. These numbers are a guide only, and can be varied depending on climatic zone. The fish that thrive in similar waters in your area are a good guide in selecting the most appropriate species for your dam.

9.1.3 Feeding

In correctly stocked dams, the natural food supply should be enough to support the fish population. There are, at present, no adequately formulated artificial diets for native fish, and trout will only eat floating or sinking pellets. Once pellets reach the bottom they become, in the long term, very expensive fertiliser. However, some dam owners like to feed pellets to trout as a rapid appraisal of the numbers and size of the fish present in the dam. This approach needs to be consistently managed to obtain meaningful results, so the fish need to be fed small amounts at the same time and place.

9.1.4 Aquatic vegetation

Vegetation may be a food source for some of the organisms on which fish feed. It can also provide shelter for both food organisms and fish. Most dams will develop aquatic vegetation naturally, but shallow dams can be completely taken over. Generally, no more than a third of the dam's surface should be covered by vegetation. When this is exceeded, control by physical removal is effective, but care must be taken not to kill fish.

9.1.5 Fish loss by escaping

Migration of rainbow trout, golden perch and silver perch is more likely than other species. If a dam overflows then it is very likely to lose fish, particularly once they reach sexual maturity. It is, therefore, worthwhile to install an upward-sloping screen, which extends well above maximum water level at the dam overflow, and sloped in the direction of the water flow. This allows debris to be pushed up onto the screen for regular cleaning off, or over the top, but prevents fish from escaping. It is necessary to ensure that the size screen apertures are small enough to prevent the smallest fish from passing through, that is, screen apertures one-third the size of the cross-section of the fish's body. Screens should also be installed around all outlet pipes used to supply water for other purposes such as irrigation or stock.

9.1.6 Muddy water

While most bodies of water will become turbid (cloudy) when they are stirred up by wind or flowing water, they usually clear when conditions are stable. If a dam remains turbid, it is most likely because of the type of clay present in the dam base and wall. Some clay types produce what are known as colloidal particles in the water. These are very small particles, which have similar electrical charges and repel each other causing them to remain in suspension (see Section 6.3.1). There are a number of chemicals that will neutralise these charges, allowing the particles to settle (see Section 7.3).

9.1.7 Algae

Excessive algal growth can taint water, clog filters and valves, deplete oxygen levels and kill stock. Algae increase in number by the action of sunlight on nutrient-rich water. In the long term, control is achieved by reducing nutrient input. This can be achieved by:

- diverting effluent from a dairy away from the dam;
- planting vegetation that will shade the dam in summer; or
- reducing run-off and associated erosion within the catchment.

Most chemical treatment for algae will kill fish and expert advice is required before treating any water in the dam (see Section 7.4).

9.1.8 Other species

At different times, a number of species have been suggested as being suitable for culture. Currently, it would appear that one or more factors make farming of them uneconomic. Interstate or tropical species may be popular in some areas, but they are not adapted to colder climates and cannot be economically grown in some places. For example, tropical barramundi and prawns die, or grow extremely slowly under cooler temperature conditions. Artificial heating is yet to be an economic solution. Native fish that occur naturally in streams and lakes can have traits that make them unsuitable for farming. Despite much interest and effort, nobody, has developed an economic system of feeding and harvesting table-sized golden perch, silver perch, freshwater catfish and Murray cod. Redfin (English perch) carry a disease that also infects other species of fish. To ensure quarantine, the movement of redfin stocks has been limited. They also have a natural propensity to over-breed and over-stock, thereby producing under-size fish. As a final disincentive to potential fish farmers, redfin attract only a poor price at market.

Many species of freshwater crayfish are native to Australia but, while yabbies show some potential for farming., other species of freshwater crayfish are not suitable. Murray spiny crayfish, Gippsland spiny crayfish, Glenelg River spiny crayfish and the giant Tasmanian crayfish grow too slowly to be commercially attractive. It may take a decade for a crop to reach market size. The Western Australian Marron is declared a noxious fish in Victoria and the tropical red claw crayfish carries diseases.

If some Asian fish species were to escape, they could have the potential to become pests. To conserve freshwater ecology these species have also been made noxious in most States of Australia and cannot be farmed.

9.2 YABBY FARMING

In Western Australia, yabbies are raised with minimal feeding and maintenance in small dams used for stock watering (Romanowski, 1994). In some cases, the owner of the dam catches and delivers the yabbies to a processing facility, but a reasonable profit for very little effort can still be expected even if they are harvested by the reseller.

While it seems possible that yabby farming may become a viable industry, it is unlikely that the potential financial returns are as great as some promoters state. The technology is simple but yabby farming can be labour intensive and requires good fish husbandry skills. To make a profit, good business skills are also required.

Landowners that grow yabbies for sale are required in most States of Australia to have a Fish Culture Permit/Licence. Anyone contemplating the farming of yabbies should check this out with the responsible State authorities that manage fisheries and/or wildlife.

9.2.1 Types of dams for yabby production

Yabby dams can be either specially constructed ponds or located in large areas of unimproved wetland. For a given area of land, specially constructed ponds have the highest potential return but cost more to establish and maintain. Crowding yabbies also increases the likelihood of disease and slows growth.

Dam dimensions may vary markedly according to the site but ideally have a depth of 1 m, and a surface area of 0.25 ha. Long, narrow ponds are considered to be more practical than round or square ponds.

There is a difference of opinion as to whether dams should be lined with plastic, gravel and limestone or natural topsoil. Farmers using all types of linings report similar growth rates so the bottom type may not be critical. High water temperatures are

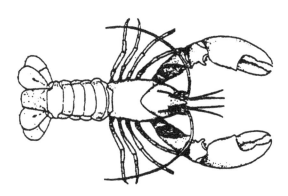

Figure 9.4 Cherax destructor (Yabby) (*Source:* DC&E, 1990).

essential for the fast growth of yabbies, so farms that are located in warm, sunny areas are likely to get the best returns.

9.2.2 Water properties

Mature aged yabbies can withstand salinities as high as 14 000 parts per million (2330 μS/cm). However, they thrive better in lower salinities. As a rule of thumb, if stock can drink the water, yabbies can survive and grow in it (see Section 7.5). Yabbies start to feed and grow when the water temperature exceeds about 13°C. As cold blooded animals, their feeding activity and growth rate increases as the temperature rises. At about 28°C, yabbies reach an optimal growth rate and higher temperatures can cause a decline in growth and death will occur when the temperature reaches 34–36°C. Increased water temperatures impose indirect impacts on yabbies through an associated decrease in dissolved oxygen and an increase in the rate of organic decomposition. Ideal water temperatures are 24–27°C.

9.2.3 Stocking and feeding

The stocking rate for yabbies is a function of farming intensity. Very high stocking rates cause problems, including fouling of water, disease and slow growth. The recommended maximum initial stocking rate is about 5 or 6 juvenile yabbies per square metre of pond bottom. As they grow, a few will die and crowding will naturally reduce.

Yabbies will eat almost anything, so it is possible to provide adequate feed at a reasonable cost. Meat and fish are acceptable, but be cautious of red meat and any processed meat that contains nitrates. Contrary to popular opinion, yabbies do not prefer rotten meat. Stock feed firms manufacture pellets specially formulated for yabbies. These are expensive, but are convenient and produce good growth. Lucerne is also a suitable food.

Yabbies will not consume excessive food, which will lie on the bottom of the pond and decompose, with consequent deterioration of water quality. To avoid this, some operators of yabby farms place food on a tray in the water which is only replenished when empty.

9.2.4 Harvesting

There are several methods for harvesting yabbies.

- The quickest and easiest method is to use a net. A disadvantage is that yabbies may be damaged in dams with irregular bottoms.
- A more expensive but popular method of harvesting is to use pots or traps.
- Another method is to drain the dam and remove the yabbies by hand. If this is done, it is best to drain the dam or pond very quickly. Lowering the water level gradually may stimulate the yabbies into burrowing, which then makes their harvesting very difficult.

9.3 NATIVE FAUNA AND TOTAL ECOSYSTEM MANAGEMENT

Native fauna can apply pressures on aquaculture farming, which, if significant, can compromise the viability of a commercial enterprise. Therefore, prior to establishment, landowners or managers should liaise with professional associations, industry contacts and people with local knowledge to develop an understanding of all the risks and uncertainties of aquaculture in the area. Using this approach, an operator will be able to avoid sites that are unsuitable, calculate possible costs, and determine the likelihood of wildlife problems. Rather than start full-scale production it is better to initiate a small-scale project. The pilot project will give an insight into problems and may enable solutions to be found before a large amount of capital has to been committed to the project. When an enterprise is considered to be potentially viable, management of the total ecosystem in the initial planning stages of development is recommended. Survival rate can be increased through strategic management.

9.3.1 Native fauna

A diversity of species of native fauna will eat fish and crayfish. Birds that consume fish and crayfish/yabbies include cormorants and darters, which are divers, and waders such as herons. Other native species that consume fish include the nocturnal, elusive water rat (*Hydromys chrysogaster*) and eels. However, it is important to note that water rats will only be present where the aquaculture enterprise is in very close proximity to a stream.

9.3.2 Managing predation by native fauna

Aquaculture has a prime focus on commercial production. This can be sustained without compromising conservation of wildlife through an holistic approach to management. Appropriate strategies can include those shown below.

- Covering ponds with netting to exclude birds from ponds. Bird netting is usually made of plastic and must give total coverage of ponds including the sides. Plastic netting can last many years and is expensive but its cost must be viewed in the context of the losses it minimises.
- Designing ponds to discourage waders, by having steep, deep margins.
- Increasing the numbers of fry slowly when first introduced to a pond.
- Including yabbies in the pond population when possible, because cormorants find them easier to catch than fish.
- Providing protective cover for fish and fry with water plants along the margins, and irregularly shaped objects, such as car bodies and tyres on the pond floor.
- Developing the adjacent areas to support alternative food sources that are adequate and attractive.

Studies have shown that more than 50 per cent of fish can be consumed in a small dam by cormorants unless alternative prey (as yabbies) are provided, and dams are stocked with few fish (around 150 per hectare) (Barlow and Bock, 1992). High stocking rates (more than 450 per hectare) are associated with high predation rates. Ideally, management should combine a number of strategies simultaneously.

All native bird and mammal predators of fish and crayfish are protected wildlife in most States of Australia. People found killing protected wildlife face fines and the possibility of jail terms. Where it can be demonstrated that wildlife damage to fish and crayfish is causing economic hardship, application can be made to obtain a licence to cull from the appropriate State agency. The application process includes field inspection and consultation with agency staff. Licences are viewed as a short term remedy, and State agencies support landowners with advice on alternative, longer term solutions.

9.4 LICENSING PROCESS

Aquaculture requires approval from several authorities to help safeguard the water resource and existing water users' entitlements. This section outlines some of the steps required to obtain a licence from a responsible authority to take and use water for such purposes.

9.4.1 Licences required

In most States, if you are interested in establishing an aquaculture enterprise or fish farm (SRW, 1995), you will need to obtain licences or approvals from the following authorities:

a Rural Water Authority (RWA)

You will need a licence from your nearest RWA office, as most aquaculture enterprises require a constant supply of fresh water. If you need to divert water from a nearby waterway you must obtain a Diversion Licence from the responsible authority, as stated under the legislation of the State. A separate licence may also be required for any necessary in-stream works (Lewis and Branson, 1996; Lewis and Beavis, 2001).

RWA licence requirements

In order to obtain a RWA diversion licence you will need to supply the information shown below.

- The proposed off-take works (including an appropriate method for measuring and monitoring water flows).
- The proposed discharge works (including an appropriate method of measuring and monitoring flows).
- The farm layout showing the number of ponds and settling ponds, and their capacities (including levels for bed and crest of ponds and internal channels).
- Surveyed cross-sections of the river and floodplain at the off-take, ponds and discharge points. The survey should also indicate on a locality plan the edge of the flood plain (that is, the edge of the high land) and any known recorded flood levels or other flooding information.
- Stream distance (in metres) between off-take and discharge points.
- Maximum diversion (in megalitres per day) required for the farm's operation.

Based on this information, the RWA will make an evaluation, taking into account water availability in the stream, off-take and discharge structures required, the likely effect on flooding and any possible detrimental effects to the water quality of local waterways.

A diversion licence will be issued on the basis of 'non-consumptive use' of the water. This means that all flows that are diverted must be returned to the stream. Charges will be determined according to current rates and the volume of water diverted.

b Environment Protection Authority (EPA)

You will need to obtain a waste water discharge licence from your local EPA office, to enable you to discharge effluent from your aquaculture farm. A dual process may exist between the RWA and EPA. However, you will need to obtain EPA approval before the RWA will issue a licence. To save time, make sure you have estimated your expected water volumes and other criteria listed below before you apply for any approval or licence.

c EPA Waste Water Discharge Licence

When issuing a waste water discharge licence, the EPA sets conditions for monitoring water quality and minimum acceptable levels for the waste water. Approvals for the volume of water discharged are based on the ratio of water diverted to water passed downstream to ensure adequate dilution flow in the stream. The EPA recommends a minimum dilution ratio, so that only a certain percentage of flows may be available for diversion.

d Other authorities

Other approvals in regard to the location and operation of a commercial enterprise will be needed from relevant authorities including the local shire council, conservation authorities, and any river management authorities. Once all these authorities have been contacted, you must arrange a joint inspection of your proposed site.

9.4.2 Other steps that may need to be taken

In addition to the permits and licences already described, it may be necessary to obtain the following:

i   A planning permit from the local shire council,
ii   A fish culture licence from your local conservation or wildlife authority,
iii   An occupancy licence from the appropriate authority, if any part of the proposed works is to be located on Crown Land,
iv   Approval for in-stream works from any local river management authority, if such an authority exists in your area (check with your shire office).

Legal

This section is not meant to be an interpretation of the law or State and Federal legislation in Australia. It has been added to make small dam owners, designers and contractors aware of some issues that may create future problems in their own country.

Liabilities that are determined in response to a dam failure affect society, governments and dam owners. The determination of the liability is the legal mechanism developed by society to recover damages due to a 'wrong' (in this case, lack of dam safety) and is another aspect of the dam safety problem. A better awareness of this legal process may help dam owners determine the steps to be taken to reduce their liability.

Asset and third party insurance is considered in some quarters to be a fundamental need and a prime responsibility of all dam owners. Therefore, the purchase of an insurance policy may be the most important measure a dam owner can take to reduce the consequences of dam failure.

10.1 LEGAL AND POLICY ASPECTS IN AUSTRALIA

Assessment of the environmental impact of small dams needs to consider not only the scientific, but also the legal/policy framework within which farmers manage their water resources. Examination of this broader decision-making context reveals current legislation that impacts upon small dam management.

First, each State or Territory has a *Water Act*, which controls the utilisation of water from streams including the licensing of on-waterway dams. Upon examination, an overall lack of specificity is a common characteristic of these Acts. Unlicensed dams are not subject to formal allocation processes. Moreover, provisions relating to on-waterway dams rely on definitions of what constitutes a waterway, stream or river, which can vary between States and are open to wide variation in interpretation.

Second, a series of Acts and policies exist which impact on small dam management and which vary between States and Territories. Table 10.1 provides an overview of key legislation and policies that have been made in relation to the construction and management of small dams at a Federal, State and Local Government level.

The legislation and policies that exist are specific in the level of control on further development and the environmental requirements during design and siting. Increasingly, farmers cannot construct dams in an ad hoc manner, particularly in rural residential areas, and where environmental considerations are enforced. However,

Table 10.1 Key Federal, State and Local Government legislation on small dams.

	Act/Policy/ Program	*Impact on small dam management*
FEDERAL		
	Income Tax Assessment Act, 1936	Provides incentives for dam development through tax write-offs for investment in bona fide storages, and drought investment allowances.
	National Landcare Program (NLP)	Grants provide proportion of funds for projects, including dams for drought proofing and erosion control.
STATE		
Tasmania	*Water Act*, 1983	Licenses on-stream dams and controls water diversion and use.
	Environmental Management and Pollution Control Act, 1996	Environmental Impact Assessment required for large on-stream dams.
	Small Dam Working Group	Assessment of licence applications on basis of impacts on fisheries and ecosystems.
	Farm Dam Development Plan	Provision of financial incentives to farmers (low interest loans and subsidies for consultants' advice).
Victoria	*Water Act*, 1989	Right to flow of water, applications for licences and authority, and water allocation.
	Special Water Supply Catchment Areas	Approval for small dams within Special WSC Areas required from Dept Natural Resources and Environment
South Australia	*Water Resources Act*, 1990 (currently being revised as Water Resources Bill, 1996)	Riparian rights, applications for licences for dams on proclaimed waterways, and water allocation.
	Development Act	Development authorisation required for specific dam applications according to size and location (including 'local heritage places', waterway and flood-zones, and flood plains) .
	Waterworks Act	Consent of the Minister is required for dams in the Adelaide Hills catchment.
	Draft Surface Water Allocation Policy	Addresses environmental issues relating to small dam development. Will include a Surface Water Management Plan.
Queensland	*Water Act*, 2000	Replaces the 1994 Dam Safety Management Guidelines
	Environmental Protection Act, 1994	Applies to dams containing 'hazardous waste' such as tailing dams and contaminated water containment dams
	Water Resources Act, 1989	Relates to permits that originated from existing dams, waterworks licences
Western Australia	*Water Act, date?*	Right of flow of water, applications for licences and allocation of water.
	Farm Water Plan	Farmers are encouraged to become self reliant with their farm water supply, in the context of the development of a water pipeline system.
	Farm Water Grants Scheme	Financial incentives for farm dam construction (up to $12 500).
	Environmental Protection Act, date?	Limited constraints applied to on-stream dams to preserve flora and to protect stream banks.

(*Continued*)

Table 10.1 Continued

	Act/Policy/ Program	Impact on small dam management
STATE New South Wales	*Water Act*, 1912	Applications for licences and authority, and allocation of water between users of a river system according to specified criteria.
	Farm Dams Policy, 1999	Landowners can harvest up to 10% of the mean annual run-off for their property as a harvestable right. A licence is needed for building dams to store water in excess of the harvestable right. Licence fees vary according to whether the dam will be used for stock/domestic or irrigation purposes.
	Farm Dam Safety Act, 1978	Prescribed dam as listed in the Act which come under a Dam Safety Committee. For more details look at References (DSC, 1996).
	Mining Act, 1992	Dams under this Act have the same conditions as Prescribed dams, but tend to be related only to tailings and mining operations.
	Soil Conservation Act, 1989	Subsidies for soil conservation works, including small dams as erosion control structures, are available under Section 10 of the Act.
Act	*Land (Planning and Environment) Act*, 1991	Approval required for any action, which disturbs soil and/or is within 20 m of a stream.
	ACT Rural Policy	At sale of property, landowner required to provide an environmental audit including details on dams, vegetation, erosion and any environmental problems. New owner required to provide a Property Management Agreement addressing degradation problems listed in audit and management proposals for development, including small dam construction.
LOCAL GOVERNMENT		
	Local Government Act (Vic), 1989	Approval may be required through Planning Scheme. Varies according to Shire Council.
	Local Government Act (SA), 1934	Permit required for dam building on an unproclaimed waterway.
	Local Government Act (NSW), 1993	Farmers required to submit a Local Environmental Plan (LEP) with their application to build a small dam. Varies according to Shire Council.
	Local Government Act (Qld), 1994	Local regulations may require approval for siting and design.

Source: modified with additions to Beavis and Howden, 1996.

decisions are at the discretion of local officers of State agencies and local governments without standardised procedures on a broad scale (Beavis and Howden, 1996).

This outline demonstrates that the legal and policy framework, which applies to small dam management, is variable. While there is considerable scope for interpretation by farmers on the construction, design, siting and management of a dam, there are also specific associated tax benefits, costs, and constraints, which must be considered in order to profit from such an investment. Furthermore, it is important to note that in their present form the Acts listed above do not appear to have had significant impact on the actual management of small dams.

The immediate questions that arise are those which address the relevance and effectiveness of such legislation, policies and regulations in ensuring that future small dam development meets standards set to preserve water quality and river health. Clearly, policy development needs to address the mismatch between policies for dam construction and those relating to water for the environment. This approach will optimise policy alignment and disable the current antagonistic system of incentives and constraints on small dams.

10.2 LIABILITY

Liability in specific instances very much depends upon the dam, the accident, the owner and the State jurisdiction in which the dam is located (US FEMA, 1987).

The liability of an owner, designer or contractor of a dam is considered general civil liability (tort). A tort is simply a civil wrong for which an injured party may recover damages from the responsible party. In most circumstances, simply causing damage is not a sufficient basis for the imposition of liability. An element of negligence must accompany the injury before liability is incurred. However, negligence is not a fixed concept. It has been modified and changed by legal precedence over time. In the simplest terms, it has been described as the violation of a duty to act as a reasonable and prudent person would act; a violation which directly results in damage to another.

The question of what 'duty of care' is imposed by society and what standard of reasonable care is imposed by the duty has undergone scrutiny and changed over time. In many instances, the duty to make a dam safe or the duty to ensure that one's property does not pose a danger to others, has been significantly increased.

While the concept of negligence has been broadened, changes in the limits of negligence do not directly affect dam owners because a separate basis of liability has long been imposed upon them. This standard is one of 'strict liability'. Strict liability is not based upon fault or negligence, rather it is based solely upon resulting damage, regardless of fault. Strict liability is generally applied to those activities that are deemed 'ultra-hazardous' and not capable of being rendered reasonably safe.

The situation of strict liability was first established in a case involving a dam in 1866 in England, *Fletcher v. Rylands* (L.R.1 Ex 265, 279–280). A dam was built in the vicinity of abandoned coal mines. The water from the dam found its way into the abandoned shafts and from there into active shafts and resulted in damage. Under present legal thought, the basis of liability for such an occurrence may well be negligent design because of failure to adequately investigate the surrounding circumstances at when the dam was built. In this decision, it was assumed that no one could have known the abandoned mine shafts existed and specifically decided that the owner was not negligent. Nonetheless, the court at that time, established the concept of strict liability for dam owners, and the owner of the dam was found to be liable for the escape of water from the dam, regardless of fault.

Thus, with a very limited number of exceptions, the general statement of liability for the owner or operator of a dam can relate back to a statement by Henry Manisty QC, who addressed the English Court of the Exchequer in 1866:

"A large collection of water is a thing pregnant with dangers, and it behoves anyone who makes a collection for his profit, to be aware how he may prejudice his neighbour by mismanaging it."

Since that time this was known as the *Fletcher v. Rylands* rule. This is now held by the High Court to have no place in Australian law, based on the precedent: *Burnie Port Authority vs. General Jones* (1994) HCA.

Strict liability has two exceptions:

1   Acts of God – natural occurrences over which the owner has no control. While acts of God are recognised as a defence, this is limited to those events over which the owner had no control and also which the owner could not, using available expertise, have anticipated.
2   Intentional acts of third parties – was established in the United States of America, Wyoming Supreme Court in the Wheatland case. In this case it was asserted that the dam had been damaged by saboteurs, and the Court recognised that illegal, intentional acts by third parties which the owner could not protect against or reasonably foresee were a viable defence to strict liability (FEMA, 1987).

In summary, existing law holds a dam owner to the highest standard of care. Pending legislation may limit liability in certain circumstances, but the general statement remains unchanged. The owner is liable for all damages caused by water escaping from a dam – despite the best efforts of the owner.

10.3   RESPONSIBILITY OF DAM OWNERS

Some State authorities provide information on acquiring a permit or licence for dam construction, dam 'hazard' and 'risk' categories, and liability in the event of dam failure. The information is intended only as a guide, and it is recommended that you seek legal advice if you have any concerns about your potential liability as a private dam owner. Also there are other issues that need to be considered.

•   If you build a dam without permission it can create problems at a later stage when repairs are needed, or perhaps when you are selling the land. Nobody wants to buy someone else's mistakes or legal problems.
•   Building a small dam is not simply a matter of walling off a gully or waterway. It is a task that requires time, effort and money to plan and complete successfully. However, no matter how well planned or constructed your dam is, there is always the possibility that it may fail.
•   Dam failure can injure people or livestock, damage residences or industrial buildings, railways or highways, interrupt the service of public utilities such as electricity, or even cause loss of life. Dam failure can also result in loss of income for the dam owner, due to lack of water for irrigation or stock.

- Dam owners are ultimately responsible for the safety of their dams. Under legislation, the owner of the land on which a dam is situated may be liable for all damage caused by the escape of water from the dam. Therefore, it is important to know beforehand if your insurance policy covers this situation.

10.3.1   Liabilities of dam owners

Every owner of a dam has a legal liability under various State legislation (see Table 10.1). In Victoria, owners of private dams may be liable for damage caused by the failure of a dam on a variety of different grounds. These grounds may arise under the *Water Act* 1989 (Vic) or at common law. The relevant sections of the *Water Act* 1989 are sections 16 to 20. Section 16 (1) of the *Water Act* 1989 provides that:

> *If –*
> *(a) there is a flow of water from the land of a person onto any other land; and*
> *(b) that flow is not reasonable; and the water causes –*
>> *(i)   injury to any other person; or*
>> *(ii)   damage to the property (whether real or personal) of any other person; or*
>> *(iii) any other person to suffer economic loss –*
>>   *the person who caused the flow is liable to pay damages to that other person in respect of that injury, damage or loss.*

Section 18 provides that the Act does not extinguish the common law liability of a private dam owner for damage caused by the escape of water from a dam.

The Act gives the Administrative Appeals Tribunal jurisdiction to hear complaints about common law claims for dams failure as well as claims brought under the Act itself. In some cases claims may also be brought in the Supreme Court.

10.3.2   Permission to build a dam

In most Australian States, a permit or licence to build a private small dam is required in a range of situations.

- If you are planning to build the dam on or over an existing waterway you may need a licence from your local Rural Water Authority (RWA). If you are uncertain about whether your proposed dam site is on a waterway, then contact your local RWA for clarification.
- Your local shire or council may require you to obtain a planning approval permit.
- You may also need to contact a government department if you plan to build your dam in a 'Proclaimed Catchment Area' that is, town water supply source, or if any stored water will intrude onto Crown Land road reserve.
- In some States there is a requirement to conform with legislation for dams described as 'Referable' (Victoria) or 'Prescribed' (New South Wales) dams, i.e:

  i   A wall that is five (5.0) metres or more high above ground level at the downstream end of the dam, and a capacity of 50 ML or more, or

ii   A wall that is ten (10.0) metres or more high above ground level at the downstream end of the dam, and a capacity of 20 megalitres or more. (1 megalitre = 1000 cubic metres).

For further details see DWR, 1987 Reports No. 3, 4 and 5; DWR, 1989 Report No. 43; DSC, 1996 and ANCOLD, 1998.

In New South Wales, a new Farm Dams Policy was introduced in 1999. If a landowner intends to construct a small dam, the Farm Dams Assessment Guide (accessible on the NSW Department of Land and Water Conservation (DLWC) website) provides a method to estimate 10 per cent of the mean annual run-off for the region in which the land is located (see references L&WC, 1999). This is the harvestable right, and refers to the amount of run-off generated from a property that the landowner has a right to harvest without a licence.

If a licence is required (because the landowner needs to use more than the harvestable right) for water use other than for stock or domestic purposes, the landowner needs to find out if an embargo is in place within the district or region. If so, then a licence can only be obtained by purchasing someone else's entitlement. Under these conditions, the landowner needs to contact the DLWC for advice.

If the area is not embargoed then the procedure of obtaining a licence is straightforward. Application forms, including an environmental questionnaire, from the DLWC need to be completed and submitted with a lodgement fee. Processing of the licence by the DLWC involves:

•   reviewing environmental impacts;
•   advertising the application in the local area; and
•   resolving objections, if they arise.

If approved, a fee is required, which will vary according to the intended use. The licence may have conditions applied in relation to environmental impacts and/or objections.

10.4 DAM FAILURE

The construction of small dams increased significantly from the 1960s onwards, with the greatest increases in numbers occurring in areas dominated by agricultural, pastoral and rural residential development. However, many of the older dams do not meet current design and construction standards, or have deteriorated with age. Surveys show that of the new dams that fail, the failure is due to lack of maintenance in the first few years after construction (see Section 6.3).

Owners of private dams may be liable for damage caused by the failure of their dam on a variety of grounds arising under legislation, or at common law:

a   if there is a flow of water from the land of a person onto any other land;
b   if that flow is 'not reasonable'; and
c   the water causes either injury to another person, property damage, or economic losses to another, then, the person causing the flow is liable to pay damages to the affected person.

This means that if water from your dam escapes onto your neighbour's property and causes damage or injury, then you may be liable to pay for the repair of that damage (see Section 10.2).

Most State agencies responsible for small dams have a policy to inform you of your responsibility to take 'all reasonable care' when constructing your dam, so as to safeguard downstream landowners and others against damage in the event of a dam failure.

10.4.1   Potential hazard and risk of small dams

Potentially hazardous dams are those which, due to their size and location, may threaten life or property if they failed. Engineers categorise dams as being 'high', 'significant' or 'low' hazard according to the potential effects of their failure (ANCOLD, 1983; DWR, 1989; ANCOLD, 1992; DC&NR, 1992 and DSC, 1996).

The term 'risk' is used to describe the possibility of a dam failing. A 'low risk' dam would therefore be less likely to fail, whereas a 'high risk' dam is considered more likely to fail. Classifying a dam as 'high risk' would not necessarily mean that it would fail. The classification may mean that the dam is constructed from materials that are likely to deteriorate over time, (for example, timber), or that its spillway has inadequate capacity. A qualified engineer is required to classify the hazard and risk of a small dam.

In Queensland the classification system for referable dams is based on the number of population at risk. A referable dam has a dam failure impact of 1 (if between 2 and 100 people are at risk) or 2 (if over 100 people are at risk). The classification system (including the determination of population at risk) for referable dams can be obtained from the *Guidelines for Failure Impact Assessments for Water Dams* published by the Department of Natural Resources and Mines (NR&M). These guidelines do not apply to dams containing 'hazardous waste' such as tailings dams and contaminated water containment dams. Those dams are now regulated by the Environmental Protection Agency under the *Environmental Protection Act* 1994.

10.4.2   Minimising the risk of dam failure

To minimise the risk of dam failure, it is recommended that private dam owners:

- inspect and maintain their dams regularly. Maintenance may include de-silting, checking for cracks along dam walls, opening the compensation pipe regularly.
- monitor conditions that may affect the safety of their dams, e.g. heavy rain after dry conditions.
- repair problems as soon as they arise, and ensure the repairs meet current standards.
- call in an engineer to investigate conditions that may result in partial or total failure of a dam, for example, cracks appearing on the dam wall.

10.4.3   In case of dam failure

Owners can play an important role in ensuring the safety of small dams by having proper operating procedures and adequate inspections, maintenance and safety surveillance. However, there should be a plan of action in case the dam fails or is threatening to do so.

The Emergency Action Plan should be directly related to the specific dam's structure and its immediate environment, and will depend on the dam owner's knowledge of the dam and its operation. It should be reviewed and if necessary updated annually. This is especially true for dams with a history of leakage, cracking, settlement, misalignment or erosion from wave action.

Since dam owners may be liable for damages associated with the failure of their dam, it is imperative to issue an effective and timely warning to downstream residents if a dam is about to fail.

In populated areas calls should be made first to the police or the State Emergency Service (SES) who will then warn and if necessary, evacuate downstream residents.

In rural areas, the warning will usually be given by telephone or direct contact with the nearest downstream residents. When telephone conversation is not possible, the person observing the dangerous condition may have to personally warn the nearest downstream residents, campers, and so on. Therefore, the owner should keep a listing of the nearest downstream residents and their phone numbers.

a Notification of a threatening situation

The following situations require immediate action:

- failure imminent (for example, water is rising and approaching the top of the embankment, or heavily coloured water is issuing from the embankment or foundation) – assess whether the structure may be saved with immediate remedial action – engineer required, release water, inform neighbours.
- failure in progress (for example, water is spilling over the embankment, or erosion of the embankment, spillway or foundation is occurring) – no chance to save the dam – contact neighbours, police, State Emergency Services.
- flooding expected or in progress upstream from dam site – contact neighbours, police, State Emergency Services (SES).

Notification and action should be planned in advance.

b Individuals and organisations who could be contacted

Addressing an emergency situation could require many people and organisations to be informed at the earliest opportunity. Keeping in mind that preservation of human life must always have first priority a contact list should be prepared and kept accessible and up to date. It might take the form shown in Table 10.2.

c Contact organisations for advice

There are a number of potential problems that may threaten the safety of a small dam. These can be expensive to remedy. Consequently, it is imperative that dam owners seek advice of an expert when in doubt.

Private and government organisations that may be able to provide advice or a service are:

- State Water Authorities;
- Rural Water Authorities;
  or a member of the:

Table 10.2 Emergency contact form.

Emergency contacts	Phone number
Neighbours 1. . . . . 2. . . . . 3.	
Local Organisations 1. Police 2. Shire or Council 3. State Emergency Service (SES) 4. RWA Regional Office	
State Organisations 1. State Emergency Service (SES) 2. Department of Conservation and Natural Resources 3. Rural Water Authority 4. State Roads Authority 5. State Electricity Authority	

Source: DC&NR, 1992.

- Association of Consulting Engineers, Australia, State Chapter; or
- Local Government Engineers Association.

In any particular area, the best advice can usually be obtained from a consulting engineer with local experience in design, construction, maintenance and repair of small dams. A consulting engineer in general practice will not always have an appropriate background, and local knowledge is usually the best guide to firms or individuals with proven record of experience.

10.4.4 Abandonment of small dams

To abandon a small dam which may have outlived its usefulness or economic life can sometimes mean more than simply walking away from it. When an owner chooses to abandon a small dam because it is no longer useful or is too costly to manage or rehabilitate, the dam should be made incapable of storing any water (either temporarily or permanently) which may constitute a risk to life or property. The dam owner is still responsible for ensuring the safety of residents and development downstream, and for the dam itself, while it is in the process of being abandoned. Breaching a dam is a complex operation. First, the un-breached sections must be left in a permanently stable condition. The breach must be wide enough not to impound significant quantities of water under flood conditions. Finally, the short and long-term stability of any sediment deposits within the storage area must be considered before commencement of the breaching operation.

Breaching should not be attempted while there is any water in storage, unless expert advice is first obtained. Environmental aspects downstream must also be considered prior to a dam being breached. It should be noted that the costs of rendering an abandoned dam safe (particularly against storm and flood events) can often be quite considerable. For any dam being abandoned or breached, professional engineering advice should be obtained.

10.5   DESIGNER AND EARTHMOVING CONTRACTOR(S)

The trend in recent negligence cases in Australia indicates that, in the case of extremely hazardous activities, professional engineers, statutory bodies and others are required to give a guarantee that no physical harm results from the conduct of their activities (Wensley, 1995). However, it is clear that it takes relatively extreme circumstances to require a guarantee of safety, as the case of *Cekan v. Haines* shows, with particular reference to the question of how far a government is required to provide resources to reduce a foreseeable risk of harm. In that case, the plaintiff received self-inflicted injuries while in a police cell while intoxicated. He sued for damages. The trial judge found that the Government, which conducted the establishment, did owe a duty of care to the plaintiff but that that duty had not been breached. All members of the Court of Appeal took into account the fact that it would have been very expensive to redesign the jail in question Ñ and by extension of the argument, all the jails in New South Wales – to prevent the type of injury which had occurred. They emphasised that even governments have limited resources and must make choices on the basis of the overall common good.

The proprietors and designers of large dams are not the insurers at law of those who may suffer physical or economic harm should something go wrong. Nonetheless, the boundaries of the law of negligence are being pushed wider all the time, and because of the extreme consequences which would follow a dam failure or over-topping, a very high standard of care will be imposed upon such persons (McMullan, 1995). Therefore, dam owners and designers must assess carefully the risks involved in the design and maintenance of such structures, and take particular care to ensure that risks, even if mathematically quite small, are minimised or managed.

The law of contract is also largely concerned with the enforcement of duties that one person has, by agreement, bound himself or herself to perform for the benefit of another. The law of torts (civil wrong) may also be seen to be concerned with breaches of duties. Those duties are not established by any agreement between persons but rather by the law itself. Thus there are, for example, duties not to assault another person, not to trespass on another's land, not to take another's goods, and to take care not to injure one's neighbour. Some duties are laid down by legislation; others are found in what is known as the common law, that is, the rules that judges recognised in the past as being necessary to enable society to function as harmoniously as possible and which judges still recognise.

10.6   PROPERTY INSURANCE

A dam owner can directly and indirectly influence the consequences of dam failure by purchasing insurance to spread costs from a single dam owner to others. Insurance may provide liability and asset protection and is important for dam owners. The level of insurance carried should be based on value of facilities at risk, potential downstream impacts, condition and age of the dam, likelihood of an incident occurring and the cost of available insurance. Insurance spreads risk among a large group of people and provides a measure of protection for the person or organisation owning the dam, which may be held personally liable. Types of coverage, availability and cost will vary from time to time, so it is advisable to seek professional advice when considering the purchase of insurance. Some insurance companies and brokers specialise in issues related to dam failure. Recommendations of insurers can normally be obtained from insurance industry representatives. Not only can damage and liability be covered but also the cost of business interruption and lost income. Insurance can spread, and reduce potential loss and therefore should be an accepted cost of doing business. Many people have avoided this cost and have paid severely for their short-sightedness (USDA, 1969 and FEMA, 1987).

References and suggested further reading

Other publications which have been used to compile some of the information in this book or which are relevant to large dams or dams with large capacities or catchment areas are listed below. It should be noted that they are primarily addressed to engineers with specialist expertise in dam design and construction.

REFERENCES

Albright and Wilson (1990). Technical Bulletin on 'Albrite', a range of Sodium Tripolyphosphates STPP, which is used in certain types of soils as an additive to minimise dam leaks.

ANCOLD (1983). *Guidelines on assessment of the consequences of dam failure.* Australian National Committee on Large Dams, AGPS, Canberra.

ANCOLD (1992). *Guidelines on dam safety management.* First draft, May 1992. Australian National Committee on Large Dams, AGPS, Canberra.

ANCOLD (1998). *Guidelines on assessment of the consequences of dam failure.* (July 1998), AGPS, Canberra.

ANZECC (1992). *National water quality management strategy.* Australian and New Zealand Environmental and Conservation Council.

Barlow, C. C. and Bock, K. (1984). Predation of fish in farm dams by cormorants, *Phalacrocorax* spp. *Australian Wildlife Research*, **11**: 559–566.

Beavis, S. G. and Howden, S.M. (1996). Effects of farm dams on water. National Land Care Program funded project, Bureau of Rural Sciences, unpublished report.

Beavis, S. G. and Lewis, B. (1999). The impact of farm dam development on water resources due to catchment yield. Water 99 Joint Congress Hydrology and Water Resources Symposium and Second International Conference on Water Resources and Environment Research Handbook and Proceedings, 6–8 July, 1999, Brisbane, pp. 465–469.

Beavis, S. G. and Lewis, B. (2001). The impact of farm dams on the management of water resources within the context of total catchment development. Third Australian Stream Management Conference, 27–29 August 2001, Brisbane, Queensland, pp. 23–27.

Bill, S. M. (1985). Using herbicides near water. *Water Talk No. 53*, Rural Water Commission, Victoria.

Burton, J. R. (1965). Water storage on the farm. A manual of design, construction and operation for Australian conditions. Bulletin No. 9, Volume 1, 236 pp. Water Research Foundation of Australia.

CDWR (March 1986). *Guidelines for the design and construction of small embankment dams.* Division of Safety of Dams, California Department of Water Resources; Sacramento, California, USA, 61 pp.

CETC (1995). Colloid Environment Technologies Company; Technical Bulletin and Manual on the use of Bentomat/Claymax a soil additive to minimise dam leaks.

Cowling, S. J. (1967). Water in Australia. The Australian Irrigator. How to Encourage Wildlife on an Irrigation Farm. March/April 1967, pp. 43–45.

DC&E (1990). A series of pamphlets on aquaculture in Victoria. Department of Conservation and Environment, Fisheries Management Division, Inland Fisheries Management Branch.

DC&NR (1992). *Your dam – an asset or a liability.* (Written by B. Lewis of Rural Water Commission, Victoria.) Department of Conservation and Natural Resources.

DoA SA (1977). *Storing water in farm dams.* (Written by J. A. Edwards.) *Extension Bulletin No. 29/77*, AGDEX 583. Department of Agriculture, South Australia.

DoA VIC (1978). Farm Water Supplies, Speakers' Notes presented at one day seminar at the McMillan Rural Studies Centre Drouin, 8 March 1978. Department of Agriculture Victoria.

DSC (1996). *A review of major dams, i.e. Prescribed Dams and Statutory functions under both the Dam Safety Act 1978 and Mining Act 1992 in NSW, Australia.* Dam Safety Committee, New South Wales.

DWR (1987). *Water Law Review, Water Law and the Individual Riparian Rights and Dam Safety Issues Paper.* Report Nos 3, 4 and 5. Department of Water Resources, Govt Printer, Melbourne, Victoria.

DWR (1989). *Inventory of Potentially Hazardous Dams.* Report No. 43, 1989, Department of Water Resources, Victoria, Jean Gordon Government Printer, Melbourne.

FEMA (1987). Dam Safety: An Owners Guidance Manual. July 1987, Colorado Division of Disaster Emergency Services, Federal Emergency Management Agency, USA, 117 pp.

Fietz, T. R. (1969). Water storage of the farm: soil, structural and seepage aspects. *Bulletin No. 9.* Vol 2, Parts 1 & 2. Water Research Foundation of Australia, Canberra?

Greening Australia (1990). *Trees at work – improving your farm dam.* A booklet on how trees help water quality, ecology. Greening Australia, Canberra?

Hazell, D., Cunningham, R., Lindenmayer, D., Mackey, B. and Osborne, W. (2001). Use of farm dams as frog habitat in an Australian agricultural landscape: factors affecting species richness and distribution. *Biological Conservation* (in press).

Hill, D. and Edquist, N. (1982). *Wildlife and farm dams – a guide.* Edited by Soil Conservation Authority and jointly published with Fisheries and Wildlife Division, Victoria.

Hislop, D. (1998). *Farm water.* CB Alexander Agriculture College, Tocal, NSW.

Hollick, M. (1975). The design of roaded catchments for farm water supplies. Institution of Engineers (Aust), *Civil Engineering Transactions*, parts 1, 2 and 3, pp. 83–96.

IEA (1987). *Australian rainfall and runoff, a guide to flood estimation.* Institution of Engineers, Australia, Canberra.

Ingles, O. G. (1984). A short study of dam failures in Australia, 1857 to 1983. *Civil Engineering Systems*, Vol. 1, June 1984.

Kingsford, R. T. (1992). Maned ducks and farm dams – a success story. *Emu*, 92, pp. 163–169.

L&WC (1999). *Rural production and water sharing; farm dam assessment guide, adding value to the natural assets of New South Wales.* NSW Dept of Land & Water Conservation, August 1999, Sydney.

Lewis, B. (1993). It pays to plan your dam. *Aqua*, the Rural Water Corporation Victoria, Quarterly magazine, Volume 1, Issue 6, Autumn 1993, pp. 12–13.

Lewis, B. (1995a). *Farm dam education program: training notes for RWA staff.* A course covering farm dam site evaluation, administration, design and documentation. Circulation of notes limited to Rural Water Authorities in Victoria.

Lewis, B. (1995b). Farming for the future: avoid costly farm failures. *Farming Ahead with the Kondinin Group*, Publishing in Western Australia for circulation in Australia, edition July 1995, No. 42, pp. 14–16.

Lewis, B. (2001a). Minimum energy design using grassed spillways. The Institution of Engineers, Sixth Conference on Hydraulics in Civil Engineering, Hobart, Tasmania, 28–30 November 2001.

Lewis, B. (2001b). The impact and issues of increased farm dam development on Victorian Water Resources. The Institution of Engineers, Sixth Conference on Hydraulics in Civil Engineering, Hobart, Tasmania, 28–30 November 2001.

Lewis, B. and Beavis, S. (2001). Licensing aquaculture development in Victoria. Third Australian Stream Management Conference, 27–29 August 2001, Brisbane, Queensland, pp. 373–377.

Lewis, B. and Branson, G. (1996). Licensing of aquaculture in Victoria. *Proceedings of Twenty-third Hydrology and Water Resources Symposium*, Hobart, Tasmania, May 1996, pp. 343–349.

Lewis, B. and O'Brien, T. (2001). Providing for fish passage at small instream structures. The Institution of Engineers, Sixth Conference on Hydraulics in Civil Engineering, Hobart, Tasmania, 28–30 November 2001.

Lewis, B. and Perera, S. (1997). The impact of increased farm dam development on Victorian water resources. *Proceedings of Twenty-fourth Hydrology and Water Resources Symposium*, Auckland NZ, Nov. 1997, pp. 297–303.

Lewis, B., O'Brien, T. and Perera, S. (1999). Providing fish passage for instream structures. *Proceedings of Second Stream Management Conference*, Adelaide, February 1999.

McMullan, J. (1995). *Engineers in the water industry – legal issues.* Published with the support of the Institution of Engineers Australia and Chambers & Company Solicitors and Attorneys.

Nelson, K. D. (1983). Farm water supplies: A survey. *Water Talk*, No. 49, January 1983, p. 3.

Nelson, K. D. (1968). *Let's look at those dams.* Reprinted 2601/68, from the Journal of the Department of Agriculture, (Victoria), A. C. Brooks, Government Printer, Melbourne.

Nelson, K. D. (1973). The whole trouble with dams. *The Journal of Agriculture*, pp. 391–395, Nov 1973, A. C. Brooks, Government Printer, Melbourne.

Nelson, K. D. (1985). *Design and construction of small earth dams.* Melbourne, Inkata Press Pty Ltd, Melbourne.

O'Brien T., Perera, S. and Lewis, B. (1999). Providing for fish passage at small instream structures. *Proceedings of Second International Conference on Water Resources & Environment Research*, July 99, Brisbane. Queensland.

Powling, I. J. (1990). Algae in water. *Water Talk*, No. 39 Rural Water Commission, Victoria, reprinted May 1990.

Romanowski, N. (1994). *Farming in ponds and dams – an introduction to freshwater aquaculture in Australia.* A Lothian book, Melbourne, Victoria, Aust, 212 pp.

SCA (1979). *Guidelines for minimising soil erosion and sedimentation from construction sites in Victoria.* Compiled under the guidance of R. J. Garvin, M. R. Knight and T. J. Richmond. Soil Conservation Authority. Printed by A. K. Atkinson, Government Printer, Melbourne, 279 pp.

SCA (1983). Farm water supply manual collated by B. Garrett (unpublished). Soil Conservation Authority.

SCS USDA (1967). A supplement to the Soil Classification System that was adapted by the U.S. Bureau of Reclamation. Soil Conservation Service, United States Department of Agriculture.

SCS USDA (1969). *Engineering field manual for conservation practices.* Soil Conservation Service, United States Department of Agriculture, USA.

SR&WSC (1970). *Farm water supplies.* State Rivers and Water Supply Commission practical advice on making the most of available water. State Rivers and Water Supply Commission, Victoria.

SR&WSC (1970). Series of pamphlets on water quality. State Rivers and Water Supply Commission, Victoria.

SR&WSC (1973). *Metric conversion tables* (March 1973). State Rivers and Water Supply Commission, Victoria.

SRW (1995). *Southern rural water.* A series of Farm Dam pamphlets, written by B. Lewis that provides information advice on building and maintaining a farm dam. Each pamphlet is Numbered 1 to 21, Victoria.

USDA (1969). *Engineering field manual for conservation practices.* Contains 17 Chapters. United States Department of Agriculture, Soil Conservation Service.

US-FEMA (1987). *Dam safety: an owner's guidance manual.* United States Federal Emergency Management Agency; 117pp.

WAWA (1991). *Guidelines for the design and construction of small farm dams (category 2) in the Warren Lefroy Rivers Area.* 28th Feb 1991 (Edition 1), Water Authority of Western Australia, 45 pp.

Wensley, R. (1995). Dam safety management towards 2000, legal implications. Gutteridge Haskins & Davey Pty Ltd Seminar on 16 August 1995.

FURTHER READING

ANCOLD (1976). *Guidelines for operation, maintenance and surveillance of dams.* February 1976. Australian National Committee on Large Dams.

ANCOLD (1986). *Guidelines on design floods for dams.* Australian National Committee on Large Dams.

Andrews, C. and Carrington, N. (1988). *The manual of fish health.* Salamander Books.

Bell, F. G. (1980). *Engineering geology and geotechnics news.* Butterworth, London.

Cadwallader, P. L. and Backhouse, G. N. (1983). *A guide to the freshwater fish of Victoria.* Government Printing Office on behalf of the Fisheries and Wildlife Division, Ministry for Conservation, Melbourne, Victoria, Australia.

Caterpillar (1980). *Caterpillar performance handbook*, Edition 11. Printed in USA.

Chanson, H. (1999). *Hydraulics of open channel flow.* Butterworth Heinemann, 544 pp., UK.

DWR (1989). *Water Victoria – an environmental handbook.* Department of Water Resources, Victoria.

GH&D (1987). *A report to the Department of Water Resources Victoria: farm dams in catchment study* (November 1987). Gutteridge, Haskins & Davey.

Lewis, B. (1991). Inspecting a dam with the experts. *Aqua*, the Rural Water Corporation Victoria, Quarterly magazine, Volume 1, Issue 1, Autumn 1991, pp. 10–11.

Lewis, B. and Branson, G. (1996). Licensing of Aquaculture Developments in Victoria. *Twenty-third Hydrology and Water Resources Symposium Proceedings, Hobart*, Tasmania Aust, May 1996, pp. 343–349.

Merrick, J. R. and Lambert, C. N. (1991). *The yabby, marron and redclaw.* Published by Merrick Publications.

NR&M QLD (2001). Safety management guidelines for referable dams (draft). Natural Resources and Mines, Queensland Government.

Papworth, M. and Lewis, B. (2001). Some thoughts on humans and waterways. Third Australian Stream Management Conference, 27–29 August 2001, Brisbane, Queensland, pp. 501–506.

QWRC (1980). *Farm water supplies. Design charts for: earthworks, pipes, pumps, channels, spillways, drainage, irrigation and water quality.* 89 pp. January 1980. Queensland Water Resource Commission.

QWRC (1984). *Farm water supplies design manual*, Vol. 1, edited by A. J. Horton and G. A. Jobling, January 1984. Farm Water Supplies Section, Irrigation Branch, Queensland Water Resource Commission, 126 pp.

RWC (1988). *Irrigation and drainage practice.* A text on the design practices that should be used in the management of water supplies. Published by the Rural Water Commission, Armadale, Victoria, 262 pp.

Sainty G. R. and Jacobs (1994). *A field guide to water plants in Australia.* Third Edition. CSIRO, Australia, Division of Water Resources, 327 pp.

SCA (1972). Training course no. 38 at Mildura Victoria on: Farm water reticulation. Course notes, used in the training on 26th June 1972. Limited circulation. Soil Conservation Authority.

SCA (1974). Compaction of soils. Soil engineering training course. Unpublished report. Soil Conservation Authority.

Southorn, N. (1995). *Farm water supplies: planning and installation.* Inkata Press, published by Reed International Books, Australia.

TADS (1990). *Training aids for dam safety; module: inspection of embankment dams.* Manual developed by the United States Interagency Steering Committee of Federal and States.

USDA (1980). *Guide for safety evaluation and periodic inspection of existing dams.* United States Department of Agriculture, Forest Services and Soil Conservation Service; Washington, D.C.

Williams, W. D. (1980). *Australian freshwater life.* Published by Macmillan.

A glossary of terminology

The water industry has embraced engineers, scientists, educationalists, farmers and other disciplines. A range of terminology is used to describe some of the physical and scientific processes required to deliver water to farms. While some are self explanatory or are commonly used within the water industry worldwide, others have evolved locally.

Abandonment   Indicates that the dam is no longer used and no longer stores water. It has been modified hydraulically and/or structurally to ensure complete and permanent safety to life, property and the environment. It should require no further operation, maintenance, surveillance or remedial work.

Abutment   The natural ground formation between the base of the dam and its crest. The natural material below the excavation surface and in the immediate surrounding formation above the normal river level or flood plain against which the ends of the dam are placed.

Appurtenant Works   Include, but are not limited to, such structures as spillways, either in or beside the dam and its rim; low level outlet works and water conduits such as tunnels, pipelines, either through the dam or its abutments.

Aquifer   A water bearing layer of rock or material or sediment below natural surface within which water is transmitted and can be removed by pumping.

Base of Dam   The foundation area of the lowest part of the main body of the dam, the portion excluding the abutments. The base elevation is considered to be at the lowest foundation level of a substantial section of the dam. It excludes isolated pockets of excavation which are not representative of the base extending from heel to toe.

Beaching   Rocks placed to dissipate the erosive force of waves on banks. The term can mean dumped rocks, usually on a prepared filter bed.

Bed   The part of a waterway or channel that is usually or normally covered with water when it is flowing.

Berm   A horizontal ledge on a dam wall used to help stabilise the slope.

Borrow pit   The source area for natural materials used in dam construction. This is often located within the storage area of the dam. Material from the borrow pit is used to build the embankment.

Catchment Yield   The volume of water which flows from a catchment past a given point (such as a stream gauging station) and is generally calculated on an annual basis. It comprises surface run-off and base flow (discharge from shallow and deep ground water).

Catchment   The area drained by the streams or waterways down to the point at which the dam is located.

Cohesion   Degree of bonding between individual soil particles. Loose sands and gravels are cohesion-less, while sticky clays are cohesive.

Contractor   A person employed to carry out earthmoving works from plans and specifications in the construction of a dam and other works for the storage of water.

Crest of Dam   Used to denote the top of a dam. However, the term Crest is sometimes applied to the level at which water may overflow the spillway section of the dam. Term 'Top of Dam' is preferred to denote uppermost surface of the dam.

Cut-off   An impervious barrier of material or concrete to intercept seepage flows through or beneath the structure.

Dam   A barrier constructed for storage, control and diversion purposes. A dam may be constructed across a natural waterway or on the periphery of a reservoir. When water is stored behind the dam for irrigation or other water supply purposes, the whole complex becomes a 'reservoir'.

Design Flood   The maximum flood for which the dam is designed.

Disused Dam   A dam which is no longer needed for any current, definite or essential purpose, but which is allowed to store water and discharge floods while subject to routine operation, maintenance, surveillance and remedial work as may be required.

Diversion   Any natural or artificial method or means by which part of a waterway is taken from its natural course.

Downchute   An inclined open channel through which water flow is directed. Surface may be grass, concrete or beached.

Embankment   An artificial bank built across a waterway to either protect adjacent land from inundation by flooding or to store water.

Engineer   A person who is professionally qualified and suitably experienced in relevant aspects of dam engineering to allow him to engage in some or all of the investigation, design, construction, repair and remedial work, operational, maintenance and abandonment activities.

Fetch   The free distance which the wind can travel to any point in raising waves, i.e. distance from furthermost full supply level in the direction of the wind.

Foundation   The material of the valley floor and abutments on which the dam is constructed.

Freeboard   The vertical distance between the design flood level and the top of the dam.

Full Supply Level   The level of the water when the dam is at maximum operating level, excluding times of flood discharge. When a controlled spillway is provided, it is the spillway crest level.

Groundwater   The segment of water below the natural surface of the ground at a pressure equal to or greater than atmospheric.

Height of Dam   The difference in level between the natural bed of the stream or waterway at the downstream toe of the bank. If it is not across a stream, channel or waterway it is the difference in level between the lowest elevation of the outside limit of the bank and the top of the dam. See definition 'Top of Dam'.

In some instances where a dam has a free-overflow spillway only or has a controlled spillway, it may be difficult to define the top of dam level as the normal abutment sections may not exist. In such cases the height is to be measured to the level arrived at by adding the design flood rise in water level to the level of the spillway crest, or to the full supply level.

Large Dam The minimum requirements adopted for determining whether a large dam qualifies for inclusion in the ICOLD World Register of Dams are as follows:

- All dams above 15 metres (50 feet) in height, measured from the lowest portion of the general foundation area to the top of the dam.
- Dams between 10 metres (33 feet) and 15 metres (50 feet) are included, provided they comply with at least one of the following conditions:
  (a) The length of the crest, i.e. the top of the dam, to be not less than 500 metres (1600 feet).
  (b) The capacity of the reservoir formed by the dam to be not less than 1 000 000 cubic metres (35 million cubic feet or 800 acre-feet).
  (c) The maximum flood discharge dealt with by the dam to be not less than 2000 cubic metres per second (70 000 cusecs).
  (d) If the dam has specifically difficult foundation problems.
  (e) If the dam is of unusual design.

Maintenance The routine work required to maintain existing works and systems (mechanical, electrical, hydraulic and civil) in a safe and functional condition.

Monitoring The observation and recording of data from measuring devices to check the performance and behavioural trends of a dam and appurtenant structures.

Outlet Works The combination of intake structure, screen, conduits, tunnels and gate-valve meters that permit water to be discharged under control from the dam.

Owner Any person, company or authority owning, leasing or occupying the land on which a dam is constructed or proposed to be constructed.

Permeability Property of a soil which allows the movement of water through its connecting pore spaces.

Phreatic Line An imaginary line which defines the hydrostatic pressure through homogenous or zoned embankment soils. It is used to indicate the stability of the embankment at full supply and when drawdown occurs, i.e. the storage is suddenly emptied leaving the embankment soils saturated with water.

Piping Sub-surface tunnel erosion which is caused by soil and water movement.

Plastic Limit The water content at the lower limit of the plastic state of a clay. It is the minimum water content at which a soil can be rolled into a thread without crumbling. In compaction of soils within the embankment this is an important element of construction.

Pore Pressure The pressure of air and water in the voids or pores of a soil.

Referable Dam Any artificial barrier, temporary or permanent, including appurtenant works which does or could impound, divert or control water, other liquids, silt, debris or other liquid-borne material and which:

- either is 10 metres (33 feet) or more in height and has a storage capacity of more than 20 000 cubic metres (16 acre-feet);

- or has a storage capacity of 50 000 cubic metres (40 acre-feet) or more and is higher than 5 metres (16 feet).

Remedial Work   The work required to repair, strengthen, re-construct, improve or modify an existing dam, appurtenant works, foundations, abutments or surrounding area to provide an adequate margin of safety (e.g. drainage, grouting, buttressing, post-tensioning, spillway or outlet works modification.

Reservoir   A dam or an artificial lake, pond or basin for storage, regulation and control of water, silt, debris or other liquid or liquid-carried material.

Reservoir Capacity   The total storage capacity of the reservoir or dam up to Full Supply Level (not up to flood level).

Safety Review   The review procedure for assessing the safety of a dam, comprising a detailed study of structural, hydraulic and hydrologic design aspects and of the records and reports from surveillance activities.

Seepage   The escape of reservoir water by percolation through, under or around the dam.

Shear strength   The resistance to deformation in soil, either from frictional or structural resistance of the soil grains or by cohesion between the surfaces of the soil particles.

Spillway   An open earth channel, weir, conduit, tunnel or other structure designed to allow discharges from the dam when water levels rise above the crest controlling flow down or into the spillway structure. The spillway is principally to discharge flood flows safely past a dam but may be used to release water for other purposes. The spillway may be uncontrolled (a free-overflow spillway) in which case discharge occurs when the dam rises above the crest. If a barrier is used to control the uppermost level of the dam it is referred to as a controlled spillway.

Spillway Crest   The uppermost portion of the overflow section which is below the 'Top of the Bank'.

Subgrade material   Material in cuts, fills and embankments immediately below the foundation material.

Surveillance   The continuing examination of the condition of a dam and its appurtenant structures and the review of operation, maintenance and monitoring procedures and results in order to determine whether a hazardous trend is developing or appears likely to develop.

Tailwater   The Full Supply Level of water that inundates upstream land within a storage.

Top of Dam   The elevation of the uppermost surface of the dam proper not taking into account any camber allowed for settlement or kerbs, parapets, guardrails or other structures that are not a part of the main water retaining structure. This elevation is usually the roadway or walkway or the non-overflow section of the dam.

Waterway   A general term for any stream, river, watercourse or creek. This may be flowing but not necessarily continuously, so that it may be a dry bed (depression along which water flows) periodically and also includes man-made artificial cut channels. The legal definition may vary between States.

Engineering specification for an earth-fill farm dam

Description of work

1 This specification is for the construction of an earth-fill farm dam and incidental works on the property of:

Landowner(s) Name:
Crown Allotment No:
Parish of:

2 This Specification shall include all operations necessary for the construction and completion of the works including the supply of all plant, labour, materials, tools and everything whatsoever necessary to carry out the works in accordance with Plan number_____, this Specification, and to the satisfaction of the Engineer.

Nature of work

3 Notwithstanding any description contained in the Plans or Specifications, the Contractor shall be responsible for satisfying himself as to the nature and extent of the specified works and the physical and legal conditions under which the works will be carried out, including means of access, extent of clearing, nature of material to be excavated, type and size of mechanical plant required, location and suitability of water supply for construction and testing purposes, and any other like matters affecting the construction of the works.

4 Provision shall be made for diverting water clear of the works during construction unless approval is given to the contrary by the Engineer. The Contractor shall take all necessary care to protect the works against flood conditions and is to provide, install, maintain, and operate all pumping equipment necessary for de-watering the site where required during progress of the works.

5 The Contractor shall be entirely responsible for arranging water supply to increase the soil moisture of the embankment material and for other construction purposes.

Engineer

6 The Engineer means the person appointed by the landholder to supervise the construction of the dam. The Principal requires that the Engineer should be well

experienced in the field of small dam engineering, who has academic qualifications acceptable for corporate membership of the Institution of Engineers Australia and who is considered experienced in small dam engineering.

Plans

7   The following plans (to be attached) shall form part of this Specification.

Survey marks for setting out works

8   A level bench mark has been established close to the site of the works and is shown on Plan Number_____. The reduced level of this benchmark is E.L_____, this being an arbitrary datum and shall be used as a permanent reference for levels during the construction and maintenance of the works and for calculating the accuracy of the works.

9   The Contractor must not move or disturb the benchmark in any way prior to or during the construction of the works unless prior approval is given by the Engineer.

10   The location of pegs and other reference features for the setting out of the new embankment centre-line and for the location of ancillary works are shown on Plan Number _____. The new embankment centre-line and ancillary works are to be located relative to these pegs and other reference features in accordance with the dimensions shown on the Plans.

Clearing and grubbing

11   The area to be covered by the embankment, borrow pits and incidental works, together with an area extending beyond the limits of each for a distance of seven (7) metres all round shall be cleared of all trees, scrub, stumps, roots, dead timber and rubbish and the same shall be removed from the vicinity of the work and burned or otherwise disposed of in a manner approved by the Engineer.

12   The area to be covered by the stored water outside the limits of the borrow pits shall be cleared of all scrub and rubbish. Trees shall be cut down stump high and removed from the vicinity of the work.

13   All trees near the site which are likely to damage or obstruct work in any way shall be cut down stump high or root felled as directed, and removed from the vicinity of the work.

14   All holes made by grubbing within the area to be covered by the embankment shall be filled with sound material and well compacted to finish flush with the natural surface.

Removal of topsoil for use in the embankment

15   Before construction of the cut-off trench or of any ancillary works within the embankment area, all grass growth and topsoil shall be removed from the area to be occupied by the embankment and shall be deposited clear of this area and reserved for use in completing the embankment and spillway.

16   All topsoil within the borrow pit excavation area, including the spillway, shall likewise be removed clear of the excavation and embankment areas and shall be reserved for use in completing the embankment and spillway.

Cut-off trench

17   Where directed, because of any doubt as to the nature of existing materials forming the foundation of the embankment, a cut-off trench extending downwards into a sufficiently impervious formation, shall be excavated. The cut-off trench shall extend for the full length of the embankment and beyond where shown on plan, until a continuous impervious zone is encountered extending to above top water level. The minimum width of the cut-off trench shall be three (3) metres unless otherwise indicated.

18   All water, loose soil, and rock shall be removed from the trench before backfilling commences. The cut-off trench shall be backfilled with selected earth-fill of the type specified for the embankment, and this soil shall have a moisture content and degree of compaction the same as that specified for the selected core zone.

19   Material excavated from the cut-off trench is to be used in the embankment if suitable, provided it is placed in the correct zone according to its classification. Otherwise unsuitable material shall be disposed of as directed by the Engineer.

Selection and placing of material for embankment zones

20   Suitable material from within the storage area, or from borrow pit if sufficient suitable material is not available from within the storage area, shall be used to construct the embankment. Earth-fill shall be placed longitudinally in progressive horizontal layers of uniform thickness of not more than two hundred (200) millimetres before compacting, inclusive of the depth of loose fill at the top of the preceding layer.

21   Where the earth-fill materials vary in permeability and/or moisture content, the embankment shall be constructed with an impervious core zone flanked by outer zones of semi pervious earth-fill as shown on the plans.

22   The core zone shall extend for the full length of the embankment and shall be built up and compacted to the dimensions shown on the plans. Where a cut-off trench is used the core zone shall be continuous with the compacted fill in the cut-off for the full width, of the top of the cut-off fill.

23   Where both dispersive and non-dispersive classified earth-fill materials are available, non-dispersive earth-fill shall be used in the core zone. The outer zones shall be built of the remaining classified earth-fill materials with the most permeable materials being placed in the downstream outer zone.

24   No silt, sand, stones over 75 millimetres diameter, surface soil, tree roots, organic matter, or other material, which will not compact properly, shall be used in the embankment except where shown on plans for special purposes.

25   Excavated material which, in the opinion of the Engineer, is not suitable for use in the embankment, shall be dumped in spoil banks clear of the works as directed.

26   During construction the Contractor shall maintain the embankment in a free draining condition in a manner satisfactory to the Engineer and no part of the

work shall be carried up more than 0.6 metres higher than any other part. Work shall be suspended whenever, in the opinion of the Engineer, it cannot be carried out satisfactorily owing to the fill becoming too wet.

27   When outlet pipes pass through fill material, this material shall be compacted up to a level of 0.6 m above the intended crown of the pipe. If the pipe is to be laid in undisturbed foundation material and will be less than 0.6 metres below the level of the stripped site, the laying of the pipe shall be delayed until the fill material above the pipe reaches a level at least 0.6 metres over the crown of the pipe. The trenches for the pipes will then be excavated as described in the relevant clauses of the Specification and the pipes and any protective concrete placed in position. The trench shall then be backfilled with approved material and satisfactorily compacted by pneumatic rammer or mechanical hand tampers in not more than 100 millimetres (loose) layers to the satisfaction of the Engineer until the cover over the pipe is at least 0.6 metres. The placing of backfill in trenches, when protective concrete surrounding the pipe has been placed, shall not commence until the concrete has cured for such times as may be determined by the Engineer. When no concrete surround is specified particular care shall be taken to avoid damage to the pipes or protective pipe coating.

Rollers and rolling

28   The whole of the material in the embankment shall be compacted to the approval of the Engineer by eight (8) passes of sheepsfoot rollers exerting a pressure of not less than thirty-five (35) kilopascals of projected area of tamping feet actually in contact with the ground at one time and with these feet projected for a length of not less than 175 millimetres.

29   If the Contractor desires to use a roller other than that specified in the previous clause, variation of depth of layers and/or number of passes may be required by the Engineer so as to effect the same degree of compaction.

30   The Contractor shall, at the direction of the Engineer, make a greater or lesser number of passes than the eight specified above, if, with the required moisture content, such change is necessary to obtain the desired compaction. In the event of a different roller or procedure being used, any adjustment required will be made by the Engineer to conform with the intent of this section of the Specification.

Compaction tests

31   Before the commencement of compaction and from time to time thereafter, the Contractor will receive notification from the Engineer as to the maximum dry density of the fill material as determined by the laboratory compaction tests. The desired degree of compaction is 95% of the maximum dry density as notified by the Engineer.

32   When the first layer of fill material has been spread and rolled the Engineer may, in order to carry out compaction tests, direct the Contractor to discontinue placing material over this compacted layer for a period not exceeding 24 hours.

33   The Contractor will subsequently be called upon to discontinue work, for reasons associated with testing of materials, only if other excavated material differs significantly from this tested first layer of fill.

Moisture content

34   Before commencement of compaction, and from time to time during construction of the bank, tests for moisture content will be carried out by the Engineer on the basis of the Standard Drop Test or other tests.

35   The Contractor will receive notification from the Engineer as to the optimum moisture content for compaction for each type of material as determined by the tests. The allowable tolerance of moisture content shall be plus or minus two per cent (+ or −2%) of the optimum moisture content as established by tests carried out in accordance with Australian Standard 1289 – 'Method of Testing Soils for Engineering Purposes'.

36   Material, which in the Engineer's opinion is too damp for compaction, shall be allowed to dry out to a moisture content satisfactory to the Engineer before rolling is commenced or continued, or such material shall be removed at the Contractor's expense.

37   If material is too dry for compaction, it shall be watered at the direction of the Engineer. This shall be by sprinkling in place on the earth-fill, or, if practicable, by sprinkling the material in the excavation.

Placement of topsoil cover

38   After completion of the embankment all loose uncompacted earth-fill material on the upstream and downstream batters shall be removed prior to spreading of topsoil. This may be carried out progressively during construction so as to make use of these materials in the appropriate zone of the embankment.

39   Topsoil shall be spread on the batters and on the crest of the embankment to give a uniform thickness of 150 millimetres measured after rolling with a crawler tractor. The purpose of this topsoil is to assist in the establishment of a suitable grass cover to minimise erosion caused by wind, rain and/or wave action. A suitable holding grass cover shall be established as soon as possible.

40   The finished dimensions of the embankment after spreading of topsoil shall conform to the drawings with a tolerance of 75 millimetres from the specified dimensions.

Spillway and spillway return slope

41   The spillway shall be excavated as shown on the plans, and the excavated material if classified as suitable shall be used in the embankment, and if not suitable it shall be disposed of into spoil heaps. Special care shall be taken to ensure that the excavation is level at the lip of the spillway return slope and to ensure that no spoil spills over onto that slope.

42   The spillway and spillway return slope shall be protected against erosion by planting and maintaining a suitable holding grass which shall be established as quickly as possible using fertilisers and watering if necessary.

43   To prevent natural runoff discharging over spillway batters and causing erosion, catch drains shall be constructed to collect this flow and discharge it away from the batters.

Installation of outlet pipe

44 A pipe outlet shall be placed under the embankment complete with outlet valve and cut-off provisions all in accordance with the plans.

45 The outlet pipe, after joining and before backfilling or encasing with concrete, shall be tested by the Contractor to the satisfaction of the Engineer.

46 When encasing the outlet pipe in concrete consisting of 1 part by volume of cement, 2 parts by volume of sand and 4 parts by volume of 20 millimetres (maximum) screenings, with sufficient water to give 18 MPa (minimum) concrete at 28 days, care should be exercised to see that the concrete completely surrounds the pipe and that no air pockets are formed beneath the pipe. The Contractor may support the pipe on approved concrete chairs to give the minimum thickness of one hundred (100) millimetres of concrete below the bottom of the pipe. The concrete should then be placed on one side of the pipe and vibrated under and up the other side, thereby helping to remove any air pockets, which would otherwise tend to form. Placing of concrete must commence at the lowest level of the pipe and work towards the higher level. Care must be taken to place the concrete encasing in lifts, which will not cause the pipe to 'float' in the trench.

47 When a concrete encasement of the outlet pipe is not specified in the plans, then the pipe trench shall be backfilled, when specified by the Engineer, with selected earth-fill material suitably moistened and carefully placed in 100 millimetres loose layers compacted with pneumatic or mechanical tampers.

Completion

48 The embankment and spillway shall be completed in every respect including neat trimming to correct lines and levels in accordance with this Specification and the accompanying plans. The works area shall be left in a clean and tidy condition at the completion of the work.

Metric and imperial conversion tables

Dimensional conversion table — imperial to metric and back to imperial

Imperial dimensions (Known)	Multiply by	Metric dimension (Unknown/known)	Multiply	Imperial dimensions (Unknown)
Area				
inches square (ft^2)	645.160	millimetres square (mm^2)	0.001550	inches square
feet square (ft^2)	0.092903	metres square (m^2)	10.763900	feet square
yards square (yd^2)	0.836127	metres square (m^2)	1.195990	yards square
acres (acres)	0.404685640	hectares (ha)	2.471053800	acres
miles square ($miles^2$)	2.589990	kilometres square (km^2)	0.386102	miles square
Length				
inches (in)	25.400	millimetres (mm)	0.039370100	inches
feet (ft)	0.304800	metres (m)	3.280840	feet
yards (yd)	0.914400	metres (m)	1.093600	yards
miles (ml)	1.609344	kilometres (km)	0.621371	miles
chains (ch)	20.100	metres (m)	0.049751200	chains
Volume				
feet cubed (ft^3)	0.0283168	metres cubed (m^3)	35.3147	feet cubed
yards cubed (yd^3)	0.764555	metres cubed (m^3)	1.307950	yards cubed
inches cubed (in^3)	16.387100	centimetres cubed (cm^3)	0.061023700	inches cubed
Mass				
pounds (lb)	0.453592370	kilogram (kg)	2.204620	pounds (lb)
ton (ton)	1.016046900	tonne (t)	0.984207	ton (ton)
pounds per foot (lb/ft)	1.488160	kilograms per metre (kg/m)	0.671969	pounds per foot
Flow rates				
feet cubed per second (ft^3/s)	0.028316800	metres cubed per second	35.314700	feet cubed per second (ft^3/s)
feet cubed per second (ft^3/s)	2.446575500	megalitres per day (ML/day)	0.408734600	feet cubed per second (ft^3/s)
gallons per minute (gal/min)	0.075768200	litres per second (l/sec)	13.198155	gallons per minute (gal/min)
gallons per minute (gal/min)	0.006546400	megalitres per day (Ml/d)	152.756420	gallons per minute (gal/min)
million gallons per day (mgd)	52.616782400	litres per second (l/s)	0.019005300	million gallons per day (mgd)
million gallons per day (mgd)	4.546090	megalitres per day (mgd)	0.219969	millon gallons per day (mgd)
acre feet per day (ac.ft/day)	1.233481800	megalitres per day (mgd)	0.810713	acre feet per day (ac.ft/day)
		megalitres per day (mgd)	0.011574	metres cubed per second
		metres cubed per second	86.400	megalitres per day

(Continued)

Dimensional conversion table — imperial to metric and back to imperial (Continued)

Imperial dimensions (Known)	Multiply by	Metric dimension (Unknown/known)	Multiply	Imperial dimensions (Unknown)
Force				
pound-force (lbf)	4.448220	newton (N)	0.224809	pound (lb)
		kilogram (k)	9.806650	newton (N)
		newton (N)	0.101972	kilogram (k)
pounds per foot (lb/ft)	1.488160	kilograms per metre (kg/m)	0.671969	pounds per foot (lb/ft)
tons per inch square (tonf/sqin)	15.444300	megapascals	0.064748800	tons per inch square (tonf/sqin)
pounds/inch square (lb/sqin)	6.894760	kilopascals	0.145037700	pounds/inch square (lb/sqin)
tons per foot square (tonf/sq ft)	107.252	kilopascals	0.009324	tons per foot square (tonf/sq ft)
pounds/foot square (lbf/sq ft)	47.880300	pascals (newtons per metre^2)	0.020885	pounds/foot square (lbf/sq ft)
Velocity				
feet per second (ft/s)	26.334720	kilometres per day (km/day)	0.037973	feet per second (ft/s)
Power				
horsepower (hp)	0.745700	kilowatt (kW)	1.341020	horsepower (hp)
foot pound per second (ft/lb/s)	1.355820	watt (W)	0.737562	foot pound per second (ft/lb/s)

Source: modified from SR&WSC, 1973.

Acres to hectares (approximately)

Acres (Ac)	Hectares (Ha)									
	0	1	2	3	4	5	6	7	8	9
0	0	0.4	0.8	1.2	1.6	2.0	2.4	2.8	3.2	3.6
10	4	4.5	4.9	5.3	5.7	6.1	6.7	6.9	7.3	7.7
20	8.1	8.5	8.9	9.3	9.7	10.1	10.5	10.9	11.3	11.7
30	12.1	12.5	12.9	13.4	13.8	14.2	14.6	15.0	15.4	15.8
40	16.2	16.6	17.0	17.4	17.8	18.2	18.6	19.0	19.4	19.8
50	20.2	20.6	21.0	21.4	21.9	22.3	22.7	23.1	23.5	23.9
60	24.3	24.7	25.1	25.5	25.9	26.3	26.7	27.1	27.5	27.9
70	28.3	28.7	29.1	29.5	29.9	30.4	30.8	31.2	31.6	32.0
80	32.4	32.8	33.2	33.6	34.0	34.4	34.8	35.2	35.6	36.0
90	36.4	36.8	37.2	37.6	38.0	38.4	38.8	39.3	39.6	40.1
100	40.5	40.9	41.3	41.7	42.1	42.5	42.9	43.3	43.7	44.1

Source: modified from SR&WSC, 1973.

Acre feet to megalitres (approximately)

Acre feet (Ac Ft)	Megalitres (ML)									
	0	1	2	3	4	5	6	7	8	9
0	0	1.2	2.5	3.7	4.9	6.2	7.4	8.6	9.9	11.1
10	12.3	13.6	14.8	16.0	17.3	18.5	19.7	21.0	22.2	23.4
20	24.7	25.9	27.1	28.4	29.6	30.8	32.1	33.3	34.5	35.8
30	37.0	38.2	39.5	40.7	41.9	43.2	44.4	45.6	46.9	48.1
40	49.3	50.6	51.8	53.0	54.3	55.5	56.7	58.0	59.2	60.4
50	61.7	62.9	64.1	65.4	66.6	67.8	69.1	70.3	71.5	72.8
60	74.0	75.2	76.5	77.7	78.9	80.2	81.4	82.6	83.9	85.1
70	86.3	87.6	88.8	90.0	91.3	92.5	93.7	95.0	96.2	97.5
80	98.7	99.9	101.1	102.4	103.6	104.9	106.1	107.3	108.6	109.8
90	111.0	112.3	113.5	114.7	116.0	117.2	118.4	119.7	120.9	122.1
100	123.3	124.6	125.8	127.1	128.3	129.5	130.8	132.0	133.2	134.5

Source: modified from SR&WSC, 1973.
Note:
Above tables are only approximate and have been rounded to the nearest 0.1.
For the conversion of acres to hectares, e.g.: 25 acres = 20 + 5 = 10.1 hectares. The same applies to acres feet to megalitres, for example, 72 acre feet = 70 + 2 = 88.8 megalitres.